T0213516

Lecture Notes in Artificial Intelligence 9806

Subseries of Lecture Notes in Computer Science

Hayato Ohwada · Kenichi Yoshida (Eds.)

Knowledge Management and Acquisition for Intelligent Systems

14th Pacific Rim
Knowledge Acquisition Workshop, PKAW 2016
Phuket, Thailand, August 22–23, 2016
Proceedings

 Springer

Editors
Hayato Ohwada
Tokyo University of Science
Noda, Chiba
Japan

Kenichi Yoshida
University of Tsukuba
Bunkyo, Tokyo
Japan

ISSN 0302-9743 ISSN 1611-3349 (electronic)
Lecture Notes in Artificial Intelligence
ISBN 978-3-319-42705-8 ISBN 978-3-319-42706-5 (eBook)
DOI 10.1007/978-3-319-42706-5

Library of Congress Control Number: 2016944819

LNCS Sublibrary: SL7 – Artificial Intelligence

Printed on acid-free paper

This Springer imprint is published by Springer Nature
The registered company is Springer International Publishing AG Switzerland

Preface

This volume contains the papers presented at PKAW2016: The 14th International Workshop on Knowledge Management and Acquisition for Intelligent Systems, held during August 22–23, 2016 in Phuket, Thailand, in conjunction with the 14th Pacific Rim International Conference on Artificial Intelligence (PRICAI 2016).

In recent years, unprecedented data, called big data, have become available and knowledge acquisition and learning from big data are increasing in importance. Various types of knowledge can be acquired not only from human experts but also from diverse data. Simultaneous acquisition from both data and human experts increases its importance. Multidisciplinary research including knowledge engineering, machine learning, natural language processing, human–computer interaction, and artificial intelligence is required. We invited authors to submit papers on all aspects of these area. Another important and related area is applications. Not only in the engineering field but also in the social science field (e.g., economics, social networks, and sociology), recent progress in knowledge acquisition and data engineering techniques is leading to interesting applications.

We invited submissions that present applications tested and deployed in real-life settings. These papers should address lessons learned from application development and deployment. As a result, a total of 61 papers were considered. Each paper was reviewed by at least two reviewers, of which 28 % were accepted as regular papers and 8 % as short papers. The papers were revised according to the reviewers' comments. Thus, this volume includes 16 regular papers and five short papers. We hope that these selected papers and the discussion during the workshop lead to new contributions in this research area.

The workshop co-chairs would like to thank all those who contributed to PKAW 2016, including the PKAW Program Committee and other reviewers for their support and timely review of papers and the PRICAI Organizing Committee for handling all of the administrative and local matters. Thanks to EasyChair for streamlining the whole process of producing this volume. Particular thanks to those who submitted papers, presented, and attended the workshop. We hope to see you again in 2018.

August 2016

Hayato Ohwada
Kenichi Yoshida

Organization

Honorary Chairs

Paul Compton   University of New South Wales, Australia
Hiroshi Motoda   Osaka University and AFOSR/AOARD, Japan

Workshop Co-chairs

Hayato Ohwada   Tokyo University of Science, Japan
Kenichi Yoshida   University of Tsukuba, Japan

Advisory Committee

Byeong-Ho Kang   School of Computing and Information Systems,
                           University of Tasmania, Australia
Deborah Richards   Macquarie University, Australia

Program Committee

Nathalie Aussenac-Gilles   IRIT CNRS, France
Quan Bai   Auckland University of Technology, New Zealand
Ghassan Beydoun   University of Wollongong, Australia
Ivan Bindoff   University of Tasmania, Australia
Xiongcai Cai   University of New South Wales, Australia
Aldo Gangemi   Université Paris 13 and CNR-ISTC, France
Udo Hahn   Jena University, Germany
Nobuhiro Inuzuka   Nagoya Institute of Technology, Japan
Toshihiro Kamishima   National Institute of Advanced Industrial Science
                           and Technology, Japan
Mihye Kim   Catholic University of Daegu, South Korea
Yang Sok Kim   University of Tasmania, Australia
Masahiro Kimura   Ryukoku University, Japan
Alfred Krzywicki   University of New South Wales, Australia
Setsuya Kurahashi   University of Tsukuba, Japan
Maria Lee   Shih Chien University
Kyongho Min   University of New South Wales, Australia
Toshiro Minami   Kyushu Institute of Information Sciences and Kyushu
                           University Library, Japan
Luke Mirowski   University of Tasmania, Australia
James Montgomery   University of Tasmania, Australia
Tsuyoshi Murata   Tokyo Institute of Technology, Japan

Kouzou Ohara	Aoyama Gakuin University, Japan
Tomonobu Ozaki	Nihon University, Japan
Son Bao Pham	College of Technology, VNU, Vietnam
Alun Preece	Cardiff University, UK
Ulrich Reimer	University of Applied Sciences St. Gallen, Switzerland
Kazumi Saito	University of Shizuoka, Japan
Derek Sleeman	University of Aberdeen, UK
Vojtěch Svátek	University of Economics, Prague, Czech Republic
Takao Terano	Tokyo Institute of Technology, Japan
Shuxiang Xu	University of Tasmania, Australia
Tetsuya Yoshida	Nara Women's University, Japan

Contents

Knowledge Acquisition and Applications

Short Papers

Knowledge Acquisition and Machine Learning

Abbreviation Identification in Clinical Notes with Level-wise Feature Engineering and Supervised Learning

Thi Ngoc Chau Vo[1(✉)], Tru Hoang Cao[1], and Tu Bao Ho[2,3]

[1] University of Technology, Vietnam National University,
Ho Chi Minh City, Vietnam
{Chauvtn, tru}@cse.hcmut.edu.vn
[2] Japan Advanced Institute of Science and Technology, Nomi, Japan
bao@jaist.ac.jp
[3] John von Neumann Institute, Vietnam National University,
Ho Chi Minh City, Vietnam

Abstract. Nowadays, electronic medical records get more popular and significant in medical, biomedical, and healthcare research activities. Their popularity and significance lead to a growing need for sharing and utilizing them from the outside. However, explicit noises in the shared records might hinder users in their efforts to understand and consume the records. One kind of explicit noises that has a strong impact on the readability of the records is a set of abbreviations written in free text in the records because of writing-time saving and record simplification. Therefore, automatically identifying abbreviations and replacing them with their correct long forms are necessary for enhancing their readability and further their sharability. In this paper, our work concentrates on abbreviation identification to lay the foundations for de-noising clinical text with abbreviation resolution. Our proposed solution to abbreviation identification is general, practical, simple but effective with level-wise feature engineering and a supervised learning mechanism. We do level-wise feature engineering to characterize each token that is either an abbreviation or a non-abbreviation at the token, sentence, and note levels to formulate a comprehensive vector representation in a vector space. After that, many open options can be made to build an abbreviation identifier in a supervised learning mechanism and the resulting identifier can be used for automatic abbreviation identification in clinical text of the electronic medical records. Experimental results on various real clinical note types have confirmed the effectiveness of our solution with high accuracy, precision, recall, and F-measure for abbreviation identification.

Keywords: Electronic medical record · Clinical note · Abbreviation identification · Level-wise feature engineering · Supervised learning · Word embedding

1 Introduction

In recent years, there has been a growing need for sharing and utilizing electronic medical records from the outside in medical, biomedical, and healthcare research activities. Free text in clinical notes of these electronic medical records often contains

© Springer International Publishing Switzerland 2016
H. Ohwada and K. Yoshida (Eds.): PKAW 2016, LNAI 9806, pp. 3–17, 2016.
DOI: 10.1007/978-3-319-42706-5_1

explicit noises such as spelling errors, variants of terms (acronyms, abbreviations, synonyms ...), unfinished sentences, etc. [8]. Such explicit noises in the shared records might hinder users in their efforts to understand and consume the records. One kind of explicit noises that has a strong impact on the readability of the records is a set of abbreviations written in free text in the records because of writing-time saving and record simplification. Those abbreviations might result in misinterpretation and confusion of the content in the electronic medical records as mentioned in [4]. So, automatically identifying abbreviations and replacing them with their correct long forms are significant for enhancing their readability with more clarity and sharability.

Regarding abbreviation resolution, we witnessed a number of the related works with many focuses on different tasks and purposes. As one of the first works considering medical abbreviations, [3] has provided a listing of medical abbreviations in 6 nonexclusive groups for English medical records. The result from [3] was well-known as Berman's list of abbreviations, helpful for abbreviation disambiguation in English clinical notes. Another effort in [23] has been given for normalizing abbreviations in clinical text and another one in [2] for enhancing the readability of discharge summaries. Furthermore, [21] has examined three natural language processing systems (MetaMap, MedLEE, cTAKES) to see how well these systems deal with abbreviations in English discharge summaries. The authors also suggested "accurate identification of clinical abbreviations is a challenging task". This suggestion is understandable because many abbreviations are dependent on context for their interpretation as mentioned in [12]. Indeed, [12] realized many abbreviations encountered have been commonly used but dependent on context. Thus, capturing the surrounding context of an abbreviation is important to distinguish itself from non-abbreviations in clinical text.

In this paper, we propose an effective solution to abbreviation identification in electronic medical records with level-wise feature engineering and supervised learning. Level-wise feature engineering is performed to characterize each token that is either an abbreviation or a non-abbreviation at the token, sentence, and note levels. Many aspects of each token (abbreviation or non-abbreviation) can be examined and captured to be able to discriminate between the tokens. Especially, their contexts are defined according to their surrounding neighbors in a continuous bag-of-words model introduced in [13]. As a result, a comprehensive vector representation for each token is achieved in a vector space. After that, many open options can be made to build an abbreviation identifier in a supervised learning mechanism. The resulting identifier can be used for identifying abbreviations automatically in clinical text. Experimental results on various real clinical note types have confirmed the effectiveness of our solution with high accuracy, precision, recall, and F-measure.

2 Abbreviation Identification in Electronic Medical Records with Level-wise Feature Engineering

In this section, we propose an abbreviation identification task on electronic medical records with level-wise feature engineering in the vector space. The task is defined in a broader view of the clinical note de-noising process with abbreviation resolution. It contributes to cleansing clinical texts of abbreviations, one kind of explicit noises.

2.1 De-noising Clinical Notes with Abbreviation Resolution

De-noising clinical notes with abbreviation resolution aims at replacing all abbreviations written in clinical notes with their correct long forms in order to improve the readability of these clinical notes and further enable their sharability for other medical and healthcare research activities. This abbreviation resolution consists of two phases: (1). Abbreviation Identification and (2). Abbreviation Disambiguation. The first phase needs to extract the parts from the free text in the clinical notes that are abbreviations and the second phase finds a long form corresponding to each abbreviation. As one long form might have many written abbreviations and one abbreviation might be used as a short form of many words or phrases, a correct single sense needs to be determined for an abbreviation and thus, abbreviation disambiguation is implied. These two phases are performed consecutively as shown in Fig. 1. The entire process will make clinical notes with noises (abbreviations) in their text cleansed and readable.

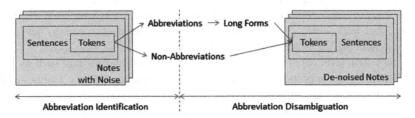

Fig. 1. De-noising clinical notes with abbreviation resolution

2.2 Abbreviation Identification Task Definition

In this subsection, we elaborate the abbreviation identification phase and define an abbreviation identification task.

As the input of this phase is a collection of clinical notes that contain free text and its output is a collection of abbreviations written in the clinical notes, we formulate an abbreviation identification task as a token-level binary classification task on the free text of the clinical notes. As well-known in data mining and machine learning, a classification task is performed in a supervised learning mechanism to classify given objects into several predefined classes. "Binary classification" means that there are two predefined classes (groups) corresponding to a group of abbreviations (class = 1) and another group of non-abbreviations (class = 0). "Token-level classification" means that each object in the classification task is defined at the token level of the clinical notes. Indeed, each object is a token obtained from the free text in the clinical notes. A token is either an abbreviation or a non-abbreviation. It is represented as a vector, called a token-level vector, so that a binary classification task can be performed in a vector space. A classification model of the binary classification task needs to be built for abbreviation identification and is called an abbreviation identifier used with the aforementioned input and producing the expected output. A token-level vector used in the phase of constructing the abbreviation identifier is called a training token-level vector. Each training vector is given a true class value which is either 1 for an abbreviation or 0 for a non-abbreviation while a vector corresponding to a token that needs to be determined as an abbreviation or not will be assigned a class value after the model classifies its vector to an appropriate class. A vector in the vector space of the task that has p dimensions is characterized by p features corresponding to p dimensions of the vector space. A feature value is of any data type in implementation depending on feature engineering we perform to represent each token. In our work, each feature value is a real number and thus, a large number of supervised learning algorithms can be utilized for the task in a vector space.

2.3   Level-wise Feature Engineering

In order to represent each token in a vector space, we first design the structure of each token in the form of a vector and then process the clinical notes to generate a corresponding vector by extracting and calculating its feature values. In our work, we do feature engineering by capturing many different aspects of each token from the most detailed level to the coarsest one. In particular, we consider the features at the token, sentence, and note levels. That is why we call our feature engineering "level-wise feature engineering". It is delineated as follows.

At the *token* level, each token is characterized by its own aspects such as word form with orthographic properties, word length, and semantics. We use 3 orthographic features named *AnyDigit*, *AnySpecialChar*, and *AllConsonants*, using 1 word length feature named *Length*, and using 2 semantic features named *inDictionary* and *isAcronym*. These token-level features are described below:

- *AnyDigit*: indicating if the current token contains any digit such as "0", "1", "2", "3", "4", "5", "6", "7", "8", and "9". If yes, one (1) is the corresponding feature value. Otherwise, zero (0) is used. The use of digits in abbreviations is little; however, they might be used to shorten the long form of some number or combined with letters in abbreviations.
- *AnySpecialChar*: indicating if the current token contains any special character such as ".", ";", ",", "-", "_", "(", ")", "@", "%", "&", and so on. If yes, one (1) is the corresponding feature value. Otherwise, zero (0) is used. It is found that abbreviations don't often contain special characters except for "-", "_", and "." for connecting the components of their long forms.
- *AllConsonants*: indicating if the current token is composed of all consonants such as "b", "c", "d", ..., "w", "x", "z". If yes, one (1) is the corresponding feature value. Otherwise, zero (0) is used. In our work, we consider acronyms to be special abbreviations. Thus, all abbreviations which are acronyms created by a sequence of the first letters of the components of their long forms tend to contain all consonants. Nonetheless, there exist some abbreviations including vowels.
- *Length*: the number of characters in the current token. It is found that most abbreviations are short due to time saving, the main purpose of abbreviation writing.
- *inDictionary*: indicating if the current token is included in a given medical dictionary. This dictionary is regarded as an external resource to provide us with the

semantics of the tokens in case they are medical terms in the bio-medical domain. If yes, one (1) is the corresponding feature value. Otherwise, zero (0) is used. This token-level feature helps realizing that the current token might be a non-abbreviation as medical terms in the dictionary are in their full forms.

- *isAcronym*: indicating if the current token matches any acronym of a medical term in the aforementioned dictionary. If yes, one (1) is the corresponding feature value. Otherwise, zero (0) is used. In contrast to *inDictionary*, this token-level feature helps us point out that the current token might be an abbreviation.

At the **sentence** level, many contextual features are defined from the surrounding words of each token in the sentence where it is contained. Different from the existing work [26] that used word forms of the surrounding words and [22] that used the characteristics of the previous/next word of each current word for capturing the context of the current word, we use a word embedding vector to encode the context of each token in a vector space. As introduced in [13], each word can be represented as a continuous vector in a vector space using either a continuous bag-of-words model or a continuous skip-gram model. The previous model predicts the current word based on the context words while the latter predicts the words in a certain range before and after the current word which is now an input. Due to the nature of our abbreviation identification task, which concentrates on deciding if a current word (token) is an abbreviation or not, we would like to capture the context of each token based on its surrounding words and thus in some certain contexts, long forms are not written, i.e. abbreviations are preferred and written. Therefore, a continuous bag-of-words model is an appropriate choice of focusing on the current token and generating the sentence-level contextual features. The number of the resulting sentence-level contextual features is the output layer size V of the continuous bag-of-words model.

At the **note** level, occurrence of each token in clinical notes is considered as a note-level feature. We use a term frequency to capture the number of occurrences of each token. Our feature engineering does not compute the percentage as we plan to perform our abbreviation resolution over time. As of that moment, updating term frequencies which are the number of occurrences is simpler than updating the percentage of each token because only the tokens in the incremental part need to be checked.

As a result, a token is represented in the following form of a vector:

$$X = \left(x_1^t, \ldots, x_{tp}^t, x_1^s, \ldots, x_{sp}^s, x_1^n, \ldots, x_{np}^n \right)$$

in a vector space of $(tp + sp + np)$ dimensions where X_i^t is a feature value of the i-th feature at the token level, X_j^s a feature value of the j-th feature at the sentence level, and X_k^n a feature value of the k-th feature at the note level; and tp is the number of token-level features, sp the number of sentence-level features, and np the number of note-level features. For our encodings in the abbreviation identification task, we design each token representation in a $(7 + V)$-dimension space with $tp = 6$, $sp = V$, and $np = 1$ where V is sp, an output layer size in the continuous bag-of-words model.

2.4   Discussion

With the abbreviation identification task definition and level-wise feature engineering, the advantages of our level-wise feature engineering are highlighted:

Firstly, unlike the related works [14, 23, 24] where feature engineering is supervised with class information in terms of "target abbreviation", ours does level-wise feature engineering in an unsupervised manner with no class information. Thus, our approach is more practical and applicable for abbreviation identification in abbreviation resolution to the coming electronic medical records over time.

Secondly, our work has defined a comprehensive representation of each token in clinical notes that can capture many different aspects of each token from the most detailed level to the roughest one, suitable for the context of abbreviation usage where an abbreviation writing habit is formed, agreed, and maintained in a group of people. Thus, sentence- and note-level features get important for abbreviation determination while token-level features are remained to characterize each token.

Thirdly, there is no restriction on abbreviation writing styles as our feature engineering does not make use of abbreviation writing styles. Above all, our level-wise feature engineering has no restriction on note structures as only token occurrence is examined at the note level. Specific note structures were not encoded into the resulting features so that many various clinical types could be supported over time. Especially, we simply consider two groups of tokens: one for abbreviations and another one for non-abbreviations. This means that there is no distinguishing between abbreviations and acronyms, leading to the capability to identify a large set of so-called short forms, i.e. abbreviations, in clinical notes.

However, we would like to note that each group of the features obtained at each level in our level-wise feature engineering is not automatically and specifically selected for any given collection of clinical notes. We choose and design them level by level based on the nature of the abbreviation resolution task, our heuristic rules, and the existing features used in the related works [22–24, 26] mentioned above.

3   An Abbreviation Identification Process Using a Supervised Learning Mechanism on Electronic Medical Records

Based on the abbreviation identification task and level-wise feature engineering defined previously, an abbreviation identification process using a supervised learning mechanism on electronic medical records is proposed. Sketched in Fig. 2, our abbreviation identification process is conducted with three parts executed sequentially: A. Data Preparation; B. Identifier Construction; and C. Abbreviation Identification.

A. Data Preparation

As we transform an abbreviation identification task on free text in clinical notes into a classification task on token-level vectors in a vector space, data preparation plays an important role to generate appropriate token-level vectors from the given clinical notes. All clinical notes need to be gone through natural language processing as they contain free text. In order to gather a collection of tokens in these clinical notes, tokenization is

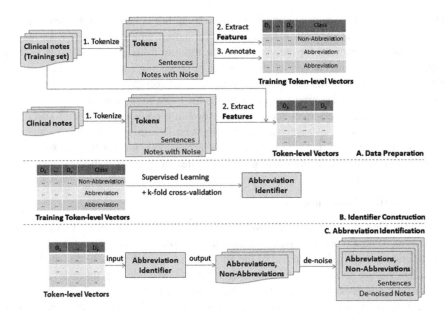

Fig. 2. The proposed abbreviation identification process using a supervised learning mechanism on electronic medical records

performed. For each resulting token, a vector is created as introduced earlier. In addition to tokenization, we use our proposed level-wise feature engineering for both training clinical notes and clinical notes that need abbreviation resolution. For both training token-level vectors and other token-level vectors, all token-level feature values are achieved after characterizing each token. All sentence- and note-level feature values are derived in an unsupervised approach after a continuous bag-of-words model is built and term frequencies are obtained using all clinical notes. An additional work is done for the tokens in the training set in order to get their class values by annotation. Annotating each token as either an abbreviation or a non-abbreviation is time-consuming but inevitable for a classification task. A true class value which is 1 or 0 is assigned to a true abbreviation or non-abbreviation in the training set as our classification task is a binary one.

An expected result of this part includes one collection of training token-level vectors and another collection of token-level vectors corresponding to the tokens in the clinical notes that are going to be de-noised with abbreviation resolution. Such an output will be fed into the next two parts for abbreviation identifier construction and abbreviation identification, respectively.

B. Identifier Construction

In our current work, the abbreviation identification task is carried out in a supervised learning mechanism as a classification task and thus, a collection of training token-level vectors needs to be ready for identifier construction in this part. After that, supervised learning is used to process these training vectors to return a classifier which is our abbreviation identifier. Before this abbreviation identifier is shifted to the next part for use, an evaluation is made in a k-fold cross-validation where our level-wise feature engineering is guaranteed for the vectors in the training folds and the test fold in each loop. If its performance is not satisfied for the abbreviation identification task, supervised learning algorithms along with Data Preparation part and particularly our level-wise feature engineering need examining for more improvement.

As the training token-level vectors are formed in a conventional vector space, we can make the most of a large number of supervised learning algorithms which are available to support this part for identifier construction. This implies that our solution is practical and applicable to de-noising clinical notes in electronic medical records to enhance their readability.

C. Abbreviation Identification

This part comes after the previous two parts as soon as token-level vectors are transferred from Data Preparation part and an abbreviation identifier is sent by Identifier Construction part. It will make use of the abbreviation identifier to decide which token is an abbreviation and which is not by classifying a corresponding token-level vector. The resulting abbreviations and non-abbreviations are then marked for the corresponding tokens in the clinical notes so that long forms of the significant abbreviations can be resolved in the next phase. How correctly a token is marked with abbreviation or non-abbreviation relies on how effective the abbreviation identifier is. Thus, active learning with human interaction should be considered to further check the returned abbreviations and non-abbreviations in practice in the future.

D. Discussion

Regarding a theoretical evaluation of our solution, which is processed in a supervised learning mechanism with level-wise feature engineering, we would like to discuss its strengths and weakness as follows.

The first remarkable point of the proposed solution is that it spends more effort on data preparation and less effort on both identifier construction and abbreviation identification. This leads to more open options to identifier construction for efficiency and effectiveness in a supervised learning phase. Besides, our solution can make abbreviation identification convenient and applicable to clinical text in a classification phase. All abbreviations in clinical text can be identified automatically.

Another advantage of our solution is that it is general, i.e. not specific, for languages, note structures, and note types used for clinical text from feature engineering to process implementation. As a derived advantage, our solution is simple and practical to be deployed in practice and lay the basis for abbreviation resolution.

It is important for us to emphasize only one weakness of our solution regarding the synchronization between the feature extraction of training token-level vectors and other token-level vectors that need abbreviation identification. As our level-wise feature engineering is done not only at the token level but also at the sentence and note levels, all clinical notes have to be gathered together to build a continuous bag-of-words model and term frequencies in an unsupervised manner. Whenever there are new clinical notes for abbreviation identification, the feature extraction of all token-level vectors needs to be performed simultaneously with the same continuous bag-of-words model and checking the same term frequencies so that sentence-level and note-level characteristics can be included in every token-level vector. As of that moment, a continuous bag-of-words model and term frequencies need to be re-generated. Nonetheless, such a weakness can be covered as we improve our solution soon in a semi-supervised learning mechanism over time.

4 Experimental Results

In order to evaluate the proposed solution, we present several experiments and provide discussions about their results. The experiments were conducted using the program written in Java for feature extraction, the word embedding implementation in Word2VecJava [20], and the supervised learning algorithm implementation in Weka 3 [18]. As an external resource, a hand-coded dictionary composed of 1995 medical terms in either Vietnamese or English is used in our experiments. Using this dictionary, we generated an acronym for each medical term regardless of its language. In the following, we elaborate our experiments for an evaluation using a triangle rule which is "at least three" for data, algorithms, and measures.

Clinical notes used in our experiments come from electronic medical records at Van Don Hospital in Vietnam [1], written in Vietnamese and English for medical terms. Details about clinical notes and abbreviations written in those notes are given in Table 1. Different from the related works, our work does not perform the identification task for each abbreviation as there are many distinct abbreviations in the current notes and the future ones. We also have no restriction on the minimum and maximum lengths of each abbreviation that needs to be identified. Therefore, for the identification task, we simply annotate each token as an abbreviation or a non-abbreviation. The percentage of abbreviations in each set of notes is calculated with respect to the total number of tokens in each set. Table 1 shows that there is an imbalance in our data set because the number of non-abbreviations is very high. This leads to our choice of evaluation measures which are Accuracy, Precision, Recall, and F-measure. Accuracy is used to check the overall performance of the resulting identifier whereas Precision, Recall, and F-measure are used for only the class of abbreviations to see how correctly the resulting identifier can recognize abbreviations. Furthermore, we prepare three different types of clinical notes including care notes, treatment order notes, and treatment progress notes. The note types are different from each other in the number of records, the number of sentences, the number of tokens, and the number of abbreviations.

Table 1. Details about clinical notes and abbreviations

Clinical note types	Care	Treatment order	Treatment progress
Patient#	2,000	2,000	2,000
Record#	12,100	4,175	4,175
Sentence#	8,978	39,206	13,852
Token#	52,109	325,496	138,602
Abbreviation#	3,031	24,693	7,641
Abbreviation%	5.82	7.59	5.51

Regarding supervised learning algorithms, we apply Random Forest with 100 trees, C4.5, and k-NN with k = 1 and Euclidean metric. The selected algorithms belong to a various number of the supervised learning categories because Random Forest is an ensemble method widely-used in the related works [22], C4.5 is a baseline decision tree algorithm also used in the related works [22, 25], and k-NN follows a lazy learning approach. Moreover, these three different algorithms have been studied very well in the machine learning research area and thus we do not focus on the fact that the best results are from which algorithm. Instead, we concentrate on a common response from these algorithms for the results of the task in our solution when several various feature sets were extracted at three different levels of detail.

In the following tables, experimental results are displayed. The experimental results on care notes are given in Table 2, the one on treatment order notes in Table 3, and the one on treatment progress notes in Table 4. In addition to these tables, we show the experimental results with different layer sizes for the continuous bag-of-word models using Random Forest in Table 5.

Table 2. Experimental results on care notes

Algorithm	Measure	Token-level	Token-level + Note-level	Sentence-level - 5	Combination-5	Combination with no external resource - 5
Random Forest	Accuracy	98.87	99.948	99.95	99.987	**99.989**
	Precision	96.1	99.8	**100**	**100**	**100**
	Recall	84	99.3	99.2	**99.8**	**99.8**
	F-measure	89.6	99.5	99.6	**99.9**	**99.9**
C4.5	Accuracy	98.879	99.931	99.86	99.941	**99.946**
	Precision	96.3	99.7	99.2	**99.8**	**99.8**
	Recall	83.9	99.1	98.4	**99.2**	**99.2**
	F-measure	89.7	99.4	98.8	**99.5**	**99.5**
1-NN	Accuracy	98.877	99.948	99.916	99.96	**99.964**
	Precision	96.3	**99.8**	99.3	99.7	99.7
	Recall	83.9	99.3	99.2	99.6	**99.7**
	F-measure	89.7	99.5	99.2	99.6	**99.7**

From the results in Tables 2, 3 and 4, we realize that sentence-level features with layer size = 5 can help achieving higher precision while token-level and note-level features achieving higher recall. In general, a combination of the features at all levels with layer size = 5 gets the highest accuracy in most cases, regardless of the selected supervised learning algorithm. Besides, we found little difference between using and not using any external resource. This might be because as of this moment, our hand-coded dictionary has a limited number of terms. In our opinion, semantic features are important to reflect each token and especially potential for the next phase to disambiguate the senses of each abbreviation.

Table 3. Experimental results on treatment order notes

Algorithm	Measure	Token-level	Token-level + Note-level	Sentence-level - 5	Combination-5	Combination with no external resource - 5
Random Forest	Accuracy	96.746	99.603	99.664	**99.672**	99.671
	Precision	81.1	97.8	**98.2**	**98.2**	**98.2**
	Recall	74.4	96.9	97.4	**97.5**	**97.5**
	F-measure	77.6	97.3	**97.8**	**97.8**	**97.8**
C4.5	Accuracy	96.747	99.601	99.624	**99.66**	99.659
	Precision	81.1	97.8	97.9	**98.1**	**98.1**
	Recall	74.4	96.9	97.1	**97.4**	**97.4**
	F-measure	77.6	97.3	97.5	**97.7**	**97.7**
1-NN	Accuracy	96.746	99.603	99.657	99.668	**99.669**
	Precision	81.1	97.8	98.1	**98.2**	**98.2**
	Recall	74.4	96.9	97.4	**97.5**	**97.5**
	F-measure	77.6	97.3	97.7	**97.8**	**97.8**

Table 4. Experimental results on treatment progress notes

Algorithm	Measure	Token-level	Token-level + Note-level	Sentence-level - 5	Combination-5	Combination with no external resource - 5
Random Forest	Accuracy	97.509	99.789	99.895	**99.907**	99.903
	Precision	94	98.5	**99.8**	99.5	99.4
	Recall	58.6	97.7	98.3	**98.8**	**98.8**
	F-measure	72.2	98.1	99	**99.1**	**99.1**
C4.5	Accuracy	97.51	99.789	99.782	99.864	**99.867**
	Precision	94	98.3	98.8	99	**99.1**
	Recall	58.6	97.9	97.3	**98.5**	**98.5**
	F-measure	72.2	98.1	98	98.7	**98.8**
1-NN	Accuracy	97.509	99.789	99.848	**99.888**	**99.888**
	Precision	94	98.5	98.8	**99**	**99**
	Recall	58.6	97.6	98.5	**99**	98.9
	F-measure	89.7	99.5	99.2	99.6	**99.7**

Table 5. Experimental results with different layer sizes using random forests

Note type	Layer size	Sentence-level				Combination with an external resource			
		Accuracy	Precision	Recall	F-measure	Accuracy	Precision	Recall	F-measure
Care Notes	5	99.950	**100.0**	99.2	99.6	**99.987**	**100.0**	**99.8**	**99.9**
	100	99.950	**100.0**	99.1	99.5	99.952	**100.0**	99.2	99.6
Treatment Order Notes	5	99.664	**98.2**	97.4	**97.8**	**99.672**	98.2	97.5	97.8
	100	99.665	**98.2**	97.4	**97.8**	99.665	98.2	97.4	97.8
Treatment Progress Notes	5	99.895	**99.8**	98.3	99.0	**99.907**	99.5	98.8	99.1
	100	99.897	**99.8**	98.3	99.0	99.899	**99.8**	98.4	99.1

From the results in Table 5, contextual features extracted by means of continuous bag-of-word models play an important role in correctly recognizing an abbreviation with very high precision but with acceptable recall. As we extend the vector space with other features at token and note levels, the abbreviation identifier can be enhanced with higher recall while precision appears to be remained. Hence, its final accuracy can be improved. This point is highlighted for both layer sizes which are 5 and 100 and for three different note types. Furthermore, it is worth noting that for these clinical notes, five contextual features extracted for each token are good enough for abbreviation identification as the experimental results with layer size = 5 are often better than those with layer size = 100. Nonetheless, "which layer size is suitable for abbreviation identification on which notes" is open in the future for the task as our work has not yet determined an appropriate value for this parameter automatically. With 5 contextual features at the sentence level, the cost for our constructing the abbreviation identifier gets increased a little as compared to that cost with 50 or 100 word embedding features in the existing works such as [24] and [11], respectively. Therefore, the total number of the resulting features at all levels in our Combination-5 experiments is 12, small but effective for abbreviation identification in those notes.

In short, our work has provided an effective solution to abbreviation identification that can be used with many various existing supervised learning algorithms and a comprehensive representation of each token in clinical notes. This solution has been examined on the real clinical notes and produced promising results on a consistent basis. It is then able to lay the foundations for abbreviation resolution in the next step where long forms need to be decided for each correctly identified abbreviation.

5 Related Works

For a comparison between the existing works and ours, an overall review on the related works is given in this section. We also present the rationales behind our experiments, which did not include any empirical comparison with these works.

First of all, we give a general review on the works in [2, 5–7, 9–11, 14–17, 19, 22–26] related to abbreviation resolution. Among these works, many related tasks have been considered to make some certain contribution to abbreviation resolution in clinical text and then further to noise cleaning and readability improvement on the clinical text in electronic medical records. For example, [10, 22, 25] focused on abbreviation detection, [5–7, 9, 11, 14, 15, 17, 19, 23, 24] on abbreviation disambiguation and expansion, and [11, 16, 24–26] on sense inventory construction for abbreviations. At this moment, our work concentrates on the first important phase of abbreviation resolution that is abbreviation identification. Using the abbreviations correctly identified, long forms can be determined and assigned to each abbreviation.

Secondly, we would like to discuss the reasons for not comparing the related works with ours in the experiments presented previously. As discussed earlier, our work aims to a more general solution to abbreviation identification. In contrast, a few related works were specific for dealing with some kinds of abbreviations in clinical text. For example, [10] has connected their solution to German abbreviation writing styles and [9] has paid attention to only the abbreviations that are 3 letters long. Another important point is that our work follows an unsupervised approach to level-wise feature engineering to be able to handle other unknown abbreviations in the future. Different from ours, the related works in [14, 23, 24] used a supervised approach in their feature engineering with respect to "target abbreviations". Our work also captures the context of each token at the sentence level using a continuous bag-of-words model in a vector space while [22] only uses local context based on the characteristics of the previous/next word of each current word and [26] uses word forms of the surrounding words. Besides, our work defines a comprehensive token-level vector representation with level-wise features in a vector space while many machine learning-based related works such as [14, 15, 25] are not based on a vector space model, leading to the different representations for clinical notes. Moreover, our work is based on a supervised learning mechanism for abbreviation identifier construction while several related works have made use of regular expressions [11], word lists and heuristic rules [25] for abbreviation identification. Last but not least, there is no available benchmark clinical data set for abbreviation identification in the present for empirical comparisons to be made with no bias. Each work has resolved this task using its own data set perhaps because of a high cost for data preparation, especially in machine learning-based works requiring large annotated data sets.

Based on our differences from the related works, it can be seen that our work has the merits of abbreviation identification so that a correct set of abbreviations can be further taken into consideration for finding their true long forms and de-noising clinical notes for readability and sharability enhancement. It has been proved with the high effectiveness of the proposed solution on various real clinical note types.

6   Conclusion

In this paper, we have taken into account an abbreviation identification task on electronic medical records. This task is formulated as a binary classification task to handle the first phase of abbreviation resolution to make electronic medical records more readable and sharable with long forms of their abbreviations. In our solution, we do level-wise feature engineering to represent each token in clinical notes in a vector space using several different aspects at token, sentence, and note levels corresponding to orthographic, word length, and semantic features at the token level; contextual features at the sentence level using the continuous bag-of-word model; and occurrence feature of each token in a given note set at the note level using term frequency, respectively. Many various existing supervised learning algorithms are then able to be utilized with the resulting token-level vectors to build an abbreviation identifier. We believe that a comprehensive set of level-wise features can help us distinguish instances of abbreviations from the others of non-abbreviations. Furthermore, our feature set does not rely much on external resources for semantics and the structure of each note type. In addition, it is proved with experimental results on real Vietnamese clinical data sets of three note types that our solution is really effective with very high accuracy, precision, recall, and F-measure values. This implies that abbreviation identification is tackled well for de-noising clinical notes with abbreviation resolution.

In the future, we plan to make our current solution more practical over time by using a semi-supervised learning mechanism instead of a supervised learning one. Besides, cleaning abbreviations from clinical notes by determining their correct long forms is one of our next steps to prepare electronic medical records for readability and sharability in further data analysis and knowledge discovery. Parallel processing for our solution to abbreviation resolution is also regarded to speed up the task. Finally, we will pay more attention to automatically determining parameter values in the task.

Acknowledgments. This work is funded by Vietnam National University at Ho Chi Minh City under the grant number B2016-42-01. In addition, we would like to thank John von Neumann Institute, Vietnam National University at Ho Chi Minh City, very much to provide us with a very powerful server machine to carry out the experiments. Moreover, this work was completed when the authors were working at Vietnam Institute for Advanced Study in Mathematics, Vietnam. Besides, our thanks go to Dr. Nguyen Thi Minh Huyen and her team at University of Science, Vietnam National University, Hanoi, Vietnam, for external resources used in the experiments and also to the administrative board at Van Don Hospital for their real clinical data and support.

References

1. A Set of Electronic Medical Records, Van Don Hospital, Vietnam, 24 February 2016
2. Adnan, M., Warren, J., Orr, M.: Iterative refinement of SemLink to enhance patient readability of discharge summaries. In: Grain, H., Schaper, L.K. (eds.) Health Informatics: Digital Health Service Delivery - The Future is Now!, pp. 128–134 (2013)
3. Berman, J.J.: Pathology abbreviated: a long review of short terms. Arch. Pathol. Lab. Med. **128**, 347–352 (2004)
4. Collard, B., Royal, A.: The use of abbreviations in surgical note keeping. Ann. Med. Surg. **4**, 100–102 (2015)
5. Henriksson, A., Moen, H., Skeppstedt, M., Daudaravičius, V., Duneld, M.: Synonym extraction and abbreviation expansion with ensembles of semantic spaces. J. Biomed. Semant. **5**(6), 1–25 (2014)
6. Kim, Y., Hurdle, J., Meystre, S.M.: Using UMLS lexical resources to disambiguate abbreviations in clinical text. In: AMIA Annual Symposium Proceedings, pp. 715–722 (2011)
7. Kim, J.-B., Oh, H.-S., Nam, S.-S., Myaeng, S.-H.: Using candidate exploration and ranking for abbreviation resolution in clinical documents. In: Proceedings of the 2013 International Conference on Healthcare Informatics, pp. 317–326 (2013)
8. Kim, M.-Y., Xu, Y., Zaiane, O.R., Goebel, R.: Recognition of patient-related named entities in noisy tele-health texts. ACM Trans. Intell. Syst. Technol. **6**(4), 59:1–59:23 (2015)
9. Kim, S., Yoon, J.: Link-topic model for biomedical abbreviation disambiguation. J. Biomed. Inform. **53**, 367–380 (2015)
10. Kreuzthaler, M., Schulz, S.: Detection of sentence boundaries and abbreviations in clinical narratives. BMC Med. Inform. Decis. Making **15**, 1–13 (2015)
11. Liu, Y., Ge, T., Mathews, K.S., Ji, H., McGuinness, D.L.: Exploiting task-oriented resources to learn word embeddings for clinical abbreviation expansion. In: Proceedings of the 2015 Workshop on Biomedical Natural Language Processing, pp. 92–97 (2015)
12. Long, W.J.: Parsing free text nursing notes. In: AMIA Annual Symposium Proceedings, p. 917 (2003)

13. Mikolov, T., Chen, K., Corrado, G., Dean, J.: Efficient estimation of word representations in vector space. In: Workshop Proceedings of the International Conference on Learning Representations (2013)
14. Moon, S., Berster, B.T., Xu, H., Cohen, T.: Word sense disambiguation of clinical abbreviations with hyperdimensional computing. In: AMIA Annual Symposium Proceedings, pp. 1007–1016 (2013)
15. Moon, S., McInnes, B., Melton, G.B.: Challenges and practical approaches with word sense disambiguation of acronyms and abbreviations in the clinical domain. Healthc. Inform. Res. **21**(1), 35–42 (2015)
16. Moon, S., Pakhomov, S., Liu, N., Ryan, J.O., Melton, G.M.: A sense inventory for clinical abbreviations and acronyms created using clinical notes and medical dictionary resources. J. Am. Med. Inform. Assoc. **21**, 299–307 (2014)
17. Pakhomov, S., Pedersen, T., Chute, C.G.: Abbreviation and acronym disambiguation in clinical discourse. In: AMIA Annual Symposium Proceedings, pp. 589–593 (2005)
18. Weka 3, Data Mining Software in Java. http://www.cs.waikato.ac.nz/ml/weka. Accessed on 22 February 2016
19. Wong, W., Glance, D.: Statistical semantic and clinician confidence analysis for correcting abbreviations and spelling errors in clinical progress notes. Artif. Intell. Med. **53**, 171–180 (2011)
20. Word2VecJava. https://github.com/medallia/Word2VecJava. Accessed on 22 February 2016
21. Wu, Y., Denny, J.C., Rosenbloom, S.T., Miller, R.A., Giuse, D.A., Xu, H.: A comparative study of current clinical natural language processing systems on handling abbreviations in discharge summaries. In: AMIA Annual Symposium Proceedings, pp. 997–1003 (2012)
22. Wu, Y., Rosenbloom, S.T., Denny, J.C., Miller, R.A., Mani, S., Giuse, D.A., Xu, H.: Detecting abbreviations in discharge summaries using machine learning methods. In: AMIA Annual Symposium Proceedings, pp. 1541–1549 (2011)
23. Wu, Y., Tang, B., Jiang, M., Moon, S., Denny, J.C., Xu, H.: Clinical acronym/abbreviation normalization using a hybrid approach. In: CLEF (2013)
24. Wu, Y., Xu, J., Zhang, Y., Xu, H.: Clinical abbreviation disambiguation using neural word embeddings. In: Proceedings of the 2015 Workshop on Biomedical Natural Language Processing (BioNLP 2015), pp. 171–176 (2015)
25. Xu, H., Stetson, P.D., Friedman, C.: A study of abbreviations in clinical notes. In: AMIA Annual Symposium Proceedings, pp. 822–825 (2007)
26. Xu, H., Stetson, P.D., Friedman, C.: Methods for building sense inventories of abbreviations in clinical notes. J. Am. Med. Inform. Assoc. **16**(1), 103–108 (2009)

A New Hybrid Rough Set and Soft Set Parameter Reduction Method for Spam E-Mail Classification Task

Masurah Mohamad and Ali Selamat[✉]

Software Engineering Research Group (SERG), Faculty of Computing,
Universiti Teknologi Malaysia, 81310 Skudai, Malaysia
masur480@perak.uitm.edu.my, aselamat@utm.my

Abstract. Internet users are always being attacked by spam messages, especially spam e-mails. Due to this issue, researchers had done many research works to find alternatives against the spam attacks. Different approaches, software and methods had been proposed in order to protect the Internet users from spam. This proposed work was inspired by the rough set theory, which was proven effective in handling uncertainties and large data set and also by the soft set theory which is a new emerging parameter reduction method that could overcome the limitation of rough set and fuzzy set theories in dealing with an uncertainty problem. The objective of this work was to propose a new hybrid parameter reduction method which could solve the uncertainty problem and inefficiency of parameterization tool issues which were used in the spam e-mail classification process. The experimental work had returned significant results which proved that the hybrid rough set and soft set parameter reduction method can be applied in the spam e-mail classification process that helps the classifier to classify spam e-mails effectively. As a recommendation, enhancement works on the functionality of this hybrid method shall be considered in different application fields, especially for the fields dealing with uncertainties problem and high dimension of data set.

Keywords: Spam · Rough set · Soft set · Parameter reduction · Classification · Hybrid

1 Introduction

Spam is one of the biggest problems for Internet users. Spam always appears in a form of advertisement such as products advertisement, get rich quick schemes and illegal service providers. Some of the spams may appear in short message service (sms), virus or Malware attachments and electronic mails (e-mail) messages. E-mail is one of the most effective and the cheapest ways of communication nowadays for all Internet users. The most popular medium for hackers to attack Internet users is by using e-mail. These spams do not only attack individual Internet users, but also affect the daily operations of companies and organizations [8,9].

© Springer International Publishing Switzerland 2016
H. Ohwada and K. Yoshida (Eds.): PKAW 2016, LNAI 9806, pp. 18–30, 2016.
DOI: 10.1007/978-3-319-42706-5_2

A number of filtering systems have been proposed to overcome this world wide problem. Consequently, different approaches have been introduced and implemented to develop these filtering systems. According to [9] there are two main approaches in detecting spam e-mail; the first approach is by using machine learning, while the other one is via knowledge engineering. The machine learning approach implements a filtering algorithm to classify the e-mail by learning the classification rules from the pre-defined e-mail training process. No rules are required to be specified by the authorities during the filtering process. This is contradictory to the knowledge engineering approach, where it classifies the spam e-mail via network information and Internet protocol address by using the rules that have been specified by the authorized person. This approach is not preferable by the user, since it is quite complicated and also inconvenient compared to the machine learning approach. Due to the ability of machine learning in classifying e-mails without the needs of pre-specified rules, many researchers have proposed different algorithms to enhance the performance of machine learning. Some of the algorithms that have been really beneficial and utilized are neural network, naive bayes, genetic algorithm, support vector machine, fuzzy set theory, rough set theory and decision tree [16].

Instead of concentrating on the performance of the classifier, parameter reduction process is also one of the most important phases in the classification task. In recent years, researchers are likely to integrate a few parameter reduction methods with the hope that the performance of the e-mail classification task will be more accurate, compared to the performance of a single method. Moreover, these hybrid methods were proposed for the sake of improving the weaknesses of the other previous hybrid methods in classifying spam e-mail. The examples of proposed hybrid methods are Support Vector Machine (SVM) with Artificial Immune System (AIS) [12], Gini Particle Swarm Optimization (PSO) with Support Vector Machine (SVM) [1] and negative selection algorithm and differential evolution [8]. To the best of our knowledge and with the literature studies done on the publications that were collected from 2010 until 2015, this was the first trial to implement the hybrid rough set (RS) and soft set (SS) parameter reduction method in the spam classification task.

The objective of this study is to propose a new hybrid parameter reduction method which integrates the two selected powerful mathematical theories in dealing with uncertainties and high dimensional data problems in any application fields. In addition, the performance of the proposed method in spam e-mail classification task also will be observed and evaluated. This paper is divided into 5 sections, where Sect. 1 contains the introduction of the proposed work, Sect. 2 defines the related works on hybrid methods which deal with the email classification task, while Sect. 3 briefly explains the method of the proposed work. Then, the discussion on the experimental work and its results are presented in Sects. 4 and 5 concludes the proposed work.

2   Related Works

This section generally explains the existing hybrid methods that deal with the email classification task. This section also highlights a number of existing hybrid rough set and soft set parameter reduction methods. Method hybridization is a process of integration between one method with another method. Hybridization promotes the alternative to solve the limitation of a single method in a particular process. Meanwhile, parameter reduction is one of the important processes that is usually applied in the decision making area. It is done in the pre-processing phase, which helps to reduce the volume of the data set and to solve the uncertainty problem before the other task such as classification or ranking task is executed. Based on the collected publications between 2010 until 2015, most of the publications preferred to explore and apply the fuzzy concept into the soft set theory instead of integrating the rough set theory with the soft set theory in the decision making problems. Two key points that are related to the hybrid methods have been identified in the area of decision making. The key points are: (1) several theories have been hybridized in order to maximize the functionality of the existing methods and (2) the hybridization of methods is proposed for the sake of minimizing the shortcoming of the original methods. Several new notions of the hybrid methods have been proposed and each of these methods have proven that a good solution could be obtained based on the given numerical examples. However, none of these publications focused on integrating rough set and soft set theories as a parameter reduction method in the email classification task.

2.1   Roles of Rough Set and Soft Set Theories as an Individual Parameter Reduction Method

Rough set theory can also be applied as a parameter reduction technique to remove unnecessary attributes by preserving the original information. Rough set has been proven as a good parameter reduction technique by many researchers and has been used extensively in many areas such as in medical diagnosis, decision making, image processing, economic and data analysis. Rough set has also been recommended as a tool that can effectively reduce the unnecessary attribute when it is integrated with other techniques in the decision making process. Recently, various parameter reduction techniques which are based on the rough set theory have been proposed such as DRSA, VC-DRSA and VP-DRSA [10]. Each of these techniques has their own ability and limitation. Basically, there are five steps involved in the rough set parameter reduction process [13]. The steps are as follows:

Step 1. Formation of the desired information table or information system
Step 2. Discretization of data
Step 3. Creation of the discernibility matrix by using the discerniblility matrix formula

Step 4. Construction of the discernibility function based on the discernibility matrix created in (Step 3)

Step 5. Attribute reduction set is obtained from the results of the discernibility function computation

Soft set theory is a theory that utilizes the advantages of rough set theory in handling imprecise and vague data [4]. Soft set theory allows the object to be defined without any restricted rules. In other words, to identify the membership function, adequate parameters are needed [11]. It is a mathematical tool which has been proposed by Molodtsov and it is independent from any insufficient parameterization tools that are inherited by several approaches such as rough set and fuzzy set theories [21]. Guan et al. [6] in their publication stated that soft set is a set of data which comprised of a record set, a set of parameters and a mapping set of selected parameters set from a power set of universe. A number of research works have been done to investigate the ability of soft set theory in the reduction process. As stated in [14], reduction process can be divided into two parts; parameter reduction and parameter value reduction. Most of the publications were likely to deal with parameter reduction instead of parameter value reduction. According to [22], there are four main phases to be followed in the soft set decision making framework when dealing with inadequate information problem. The first is the acquisition of incomplete information, followed by implementing the missing data filling algorithm to calculate the attribute weight according to different requirements. The third is calculating the target values for each item or object and finally, making the decision according to the final optimal weighted values.

2.2   Recent Researches on Hybrid Parameter Reduction Approach

According to Ma et al. [14], fuzzy soft sets is another soft set approach which has been introduced in 2001 to solve many problems including uncertainties. It is an extension of the classical soft sets approach which has been proposed in order to solve the decision making problems in the real world situation. There are many publications that contributed to fuzzy soft sets. One of the publications was written by Geng et al. [5] who proposed a method that provides approximate description of objects in an intuitionistic fuzzy environment and also with the additional information on weight attributes in solving multi-attribute decision making problems in 2011. Soft fuzzy rough sets is a combination of three mathematical tools; the soft set theory, fuzzy set theory and rough set theory which are almost related when dealing with uncertainties and vagueness problems. It was introduced by Feng et al. [3] who had investigated the problem and consequences of integrating these three theories that was inspired by Dubois and Prades research work and named as rough fuzzy sets. Then, Meng et al. [15] redefined the concept proposed by Feng et al.10 in 2011 by introducing the new soft approximation space that considered several issues which had risen from the previous research work.

2.3 Existing Works on Hybrid Approach in the Email Classification Task

This subsection describes a number of existing works which implemented different types of hybrid approaches in the email classification task. Some of the works proposed the integration of a few methods or theories as a parameter reduction method, while the other works proposed hybrid classifier methods or algorithms to execute the classification process. Table 1 lists some of the existing works related to the hybrid approach in the email classification task. These publications have given an inspiration to further novel works on the ability of hybrid approach, especially on the capability of the parameter reduction method to assist the classifier in the spam email classification task.

Table 1. Existing works on hybrid approach in the email classification task.

Approaches	Advantages	Disadvantages
A Hybrid Gini PSO-SVM Feature Selection Based on Taguchi Method [1]	The proposed method returned a high rate of precision value by using small number of attributes	Produced lower rate of recall percentage when compared to the other benchmark methods
Hybrid email spam detection model with negative selection algorithm and differential evolution [9]	The classification result was better than the original model (Negative selection algorithm) in detecting spam email	The proposed model might used other existing hybrid models as a comparison
	The proposed model was able to improve and optimized the process of generating the detector	
A hybrid approach for spam filtering using support vector machine and artificial immune system [12]	The proposed approach returned a better result when compared to the two original filtering techniques; support vector machine (SVM) and artificial immune system (AIS)	The proposed model might used other existing hybrid models as a comparison

3 Flow of the Propose Method

The proposed method integrated two mathematical tools; the rough set and soft set theories to serve as a parameter reduction method in the spam e-mail classification task. The aim of this method was to enhance the performance of the classifier in the classification process, especially for the spam e-mail classification problem. The aim of this proposed method was also to increase the performance of a single parameter reduction method in generating the reduction sets. There were four processes in the proposed method: (i) Pre-processing task, (ii) Parameter reduction and selection processes, (iii) New input value generation process and (iv) Classification process. The following subsections explained each of these phases.

3.1   Pre-processing Task

Before the spam e-mails were classified by the classifier, the e-mails will go through several pre-processing tasks for cleaning and formatting purposes. Similar to the other existing works by [1, 12], tokenization, stopword removal and stemming processes were used to remove the unwanted information which could affect the classification result. Next, the process of converting the e-mail document into a vector space model by using the term frequency inverse document frequency (TF-IDF) algorithm was executed. Then, the e-mails were formatted into a specific input file format according to the selected software or algorithm. After the input data had been formatted, the process of feature extraction or also known as parameter reduction and feature selection were executed. In this study, the process of feature reduction and feature selection were done in one process which was called as the hybrid parameter reduction method.

3.2   The Proposed Hybrid Parameter Reduction Method

Figure 1 demonstrates the proposed method of hybrid rough set and soft set parameter reduction process in the spam e-mail classification task. The hybrid parameter reduction method was implemented after the pre-processing task was done. It was executed after all the e-mail files had been cleaned up by using the stop word removal and word stemming functions. The process of the hybrid parameter reduction method consisted of two main phases.

Phase 1: the implementation of rough set attribute reduction method by using the exhaustive algorithm which was executed in the rough set exploration system (RSES). The exhaustive algorithm was implemented because of its outstanding performance in returning high prediction accuracy [2]. The RSES can be downloaded from this link http://www.mimuw.edu.pl/szczuka/rses/about. html. At this stage, the unnecessary attributes will be evaluated and eliminated for the first phase. Then, several attribute sets were listed as an output to the reduction process. The attribute set was selected based on the following steps:

Step 1: Select the attribute set which had the most attributed number as an input for the phase 2 process.
Step 2: Select the first attribute set which had the most attributed number if the output of phase 1 listed more than one sets of the most attributed number.

Phase 2: Basically, there were four main steps in the soft set parameter reduction process:

Step 1: Transformation of data set into a Boolean value information system
Step 2: Input the soft set data selection
Step 3: Calculate the significant weight of the reduction values
Step 4: Choose the best and most optimal reduction value

The parameter reduction algorithm used in this proposed work was introduced by Herawan et al. [7,18]. It was executed by using Matlab R2014a. The output of phase 2 will generate a number of simplified attribute sets according to the significant weight calculation. Then one of the attribute sets will be selected as an input to the classification process based on the best and most optimal reduction value. According to [7], the best and most optimal reduction value was chosen based on the attribute sets that contained the highest number of attributes. If the attribute sets contained more than one sets of the highest attribute set, the best attribute set might be randomly selected by the decision maker.

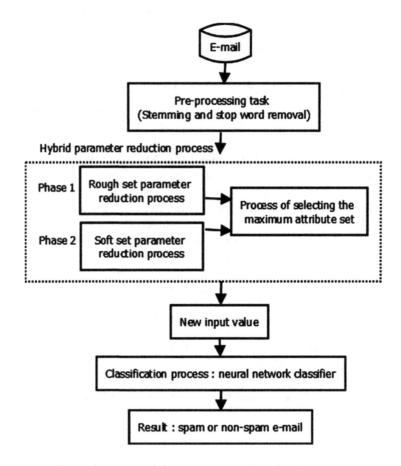

Fig. 1. Structure of the spam e-mail classification process

3.3   New Input Value Generation Process

In this phase, the input data which had went through the hybrid parameter reduction process were prepared for the classification process. As mentioned in the previous subsection, the new input data were taken from the output of the second phase in the hybrid parameter reduction process. The input data had gone through the selection process based on the steps that were stated in phase 1 in the hybrid parameter reduction process.

3.4   Classification Process

In this case study, neural pattern recognition application tool has been used to execute the classification task, while the scaled conjugate backpropagation was used to train the network. Neural network is one of the methods that may generate good classification results [17]. In this phase, the new e-mail data set will be divided into three parts; 70 % for training, 15 % for testing and another 15 % for the validation process. The network was trained several times until the desired results were obtained. Therefore, to accomplish this work, Matlab R2014a software was used to execute the classification task.

The results were evaluated by using six standard performance measures; accuracy (ACC), sensitivity (SENS), specificity (SPEC), positive predictive value (PPV), negative predictive value (NPV) and receiver operating characteristic (ROC) curves [19]. In addition, F-measure (F1) was also employed in order to observe whether the positive predictive value (PPV) and sensitivity were evenly weighted [9]. The formulations were based on TP, TN, FP and FN where TP was a true positive that presented the correctly classified message into the positives class and TN was a true negative that presented the correctly classified message into the negatives class. FP was false positive classification which represented the incorrectly classified message into the positive class, while FN was false negative that represented the incorrectly classified message into the negative class.

4   Experimental Results and Discussion

Spambase e-mail data set had been used for the experimental work and can be downloaded from the UCI Machine Learning Repository page; https://archive. ics.uci.edu/ml/datasets/Spambase. The data set consisted of 4601 emails divided into 1813 spam messages and 2788 non-spam messages. The number of attribute was 58, including the attribute of the target result. The data set also contained several missing values which were useful to test the performance of the proposed hybrid method towards the uncertainty issue. According to [9], Spambase is one of the best e-mail data set to be tested as it will return good results in the learning and testing processes. Table 2 describes the characteristics of the Spambase data set. Meanwhile, Table 3 depicts the number of simplified attributes after the hybrid parameter reduction process.

Table 2. Data set characteristics.

Item	Description
Missing values	Yes
Attribute characteristics	Integer and real
Attribute division	Number of attributes
Word type	48 continuous real [0,100]
Char type	6 continuous real [0,100]
Average length of capital letters in sequence	1 continuous real [1,...]
Longest capital letters in sequence	1 continuous integer [1,...]
Total number of capital letters in the e-mail	1 continuous integer [1,...]
Class decision	1 nominal 0,1

Table 3. Output of the hybrid parameter reduction process.

Process	Number of attributes
Pre-processing task	58
Phase 1: Rough set parameter reduction process	16
Phase 2: Soft set parameter reduction process	16
New input attribute set value	16

4.1   Evaluation Measures

In order to validate the performance of the classification results, the proposed method was compared with other two hybrid parameter reduction methods; the principle component analysis (PCA) method with rough set theory and the information gain (IG) method with rough set theory. PCA and IG were selected as comparison to the proposed hybrid method because of the performance and also because they had been widely used in the parameter reduction process [1,20]. The performance accuracy was also validated by using precision, also known as PPV, and recall, or also known as sensitivity, where these two evaluation measures are usually used in the spam classification task [1].

4.2   Results Discussion

The classification results are presented in Table 4 and Fig. 2 while Fig. 3 denotes the statistical analysis by employing the precision, recall and F-measure formula for each hybrid parameter reduction methods

As presented in Table 4 above, all the hybrid methods performed well in the classification task. The new proposed hybrid parameter reduction method also gave a competitive result in the parameter reduction process, even though it applied fewer attribute numbers than the RS with the PCA hybrid method. The attribute numbers were denoted by using the bracket stated beside the name of

Table 4. Classification results.

Hybrid methods	Overall performance %					
	ACC	SENS	SPEC	PPV	NPV	F-measure
RS + SS (16)	90.3	94.4	84.0	90.1	90.7	92.2
RS + PCA (38)	91.1	93.3	87.7	92.1	89.4	92.7
RS + IG (16)	90.3	93.5	85.4	90.8	89.5	92.13

the hybrid method, for example RS+SS (16). It helped the neural network classifier to produce a significant accuracy rate which was 90.3 % that equalled with the accuracy rate of the hybrid rough set and information gain parameter reduction method. However, these two hybrid methods had slightly different accuracy rates between the hybrid rough set and the principle component analysis parameter reduction method, where the accuracy rate increased by only 0.8 %. Even though the proposed method had the same accuracy rate with the hybrid RS and IG, the PPV percentage of the proposed method was the lowest among the three applied methods, which only achieved 90.1 % but still produced a competitive result. In terms of sensitivity, the proposed method had produced the highest result among the three methods which was 94.4 % and proved that the proposed method was able to filter the spam e-mails effectively [1]. Referring to the F-measure values, the proposed method returned the lowest percentage score which reached only 92.2 %, 0.5 % and 0.11 % difference from the other two methods. Nevertheless, the F-measure score of the proposed method still offered a good assistant to the classifier in classifying the spam e-mails.

The performance of these hybrid methods can also be referred in Fig. 3 by using ROC curves. All three methods had helped the neural network classifier to perform well in the classification task, whereas all ROC curves nearly pointed to the upper-left corner with high sensitivity and specificity values. The results

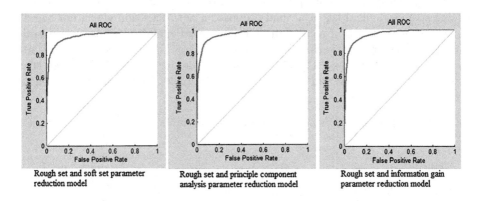

Fig. 2. ROC curves of the classification results for three hybrid parameter reduction methods

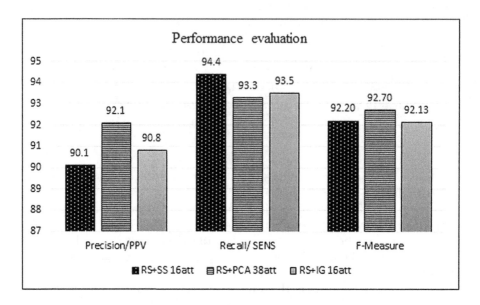

Fig. 3. Performance evaluation of hybrid parameter reduction methods

might be affected by the number of the attributes used in the classification process. This was shown by the two hybrid parameter reduction methods; (i) RS and SS and (ii) RS and IG, which produced the same number of attributes in the reduction process which was 16 attributes. Meanwhile, the RS and PCA hybrid method had produced 38 attributes to be used in the classification process. It can be concluded that, the accuracy rate will be high and improved if more attributes have been used in the classification process.

In addition, to verify the obtained results, the results of the other existing works that used the same data set were compared. The referred publications were written by [9,16], where both publications introduced different methods of identifying spam e-mails. Idris et al. had proposed a new detection algorithm based on the negative selection algorithm and differential evolution whilst Rathi and Pareek had explored the performance of the Best-First feature selection algorithm towards several classifiers. Table 5 provides the comparison results between the proposed hybrid method and the selected existing works. The results showed that

Table 5. Comparison results between the proposed hybrid method and selected existing works.

Methods	Accuracy rate %
Proposed method	90.3
Negative selection algorithm and differential evolution [9]	83.06
Best-First feature selection algorithm with Random Tree [16]	99.72

all three proposed works had performed well in the case of spam filtering, especially in the work done by [16]. It returned an outstanding result which had nearly 100 % accuracy rate. Nevertheless, the proposed hybrid method still produced a significant result, followed by the detection method proposed by [9].

5   Conclusion

One of the best ways to help the classifier to perform the best classification result is by applying the feature reduction and feature selection processes. These processes do not assist the classifier in gaining a high classification accuracy rate, but it may reduce the processing time of the classification process. Thus, motivated from this issue, a new hybrid parameter reduction method was proposed by combining two powerful mathematical theories; the rough set and soft set. To test the ability of the proposed hybrid method, a spam e-mail classification case was implemented. As demonstrated in the previous section, it was proven that the proposed hybrid method helped the classifier to generate a significant result in the spam classification process. This study also showed that the proposed method had generated a good result when compared with other existing works. Furthermore, as a recommendation, different types of data sets which contain a large number of attributes and instances should be tested to make this proposed work more beneficial to other application areas. Besides that, different types of classifiers such as the support vector machine (SVM) and k-Nearest Neighbor (kNN) were alos suggested to be applied with the proposed hybrid method to improve the classification results. We also planned to evaluate the performance of the proposed method with a number of single parameter reduction methods and to develop a dynamic hybrid parameter reduction method where the two applied methods, the rough set and soft set, are able to work in an inverse order.

References

1. Allias, N., Noor, M.N.M.M., Ismail, M.N.: A hybrid gini pso-svm feature selection based on taguchi method: an evaluation on email filtering. In: Proceedings of the 8th International Conference on Ubiquitous Information Management and Communication, p. 55. ACM (2014)
2. Chung, W., Tseng, T.L.: Discovering business intelligence from online product reviews: a rule-induction framework. Expert Syst. Appl. **39**(15), 11870–11879 (2012)
3. Feng, F., Li, C., Davvaz, B., Ali, M.I.: Soft sets combined with fuzzy sets and rough sets: a tentative approach. Soft Comput. **14**(9), 899–911 (2010)
4. Feng, F., Liu, X., Leoreanu-Fotea, V., Jun, Y.B.: Soft sets and soft rough sets. Inf. Sci. **181**(6), 1125–1137 (2011)
5. Geng, S., Li, Y., Feng, F., Wang, X.: Generalized intuitionistic fuzzy soft sets and multiattribute decision making. In: Proceedings - 2011 4th International Conference on Biomedical Engineering and Informatics, BMEI 2011, vol. 4, pp. 2206–2211 (2011)

6. Guan, X., Li, Y., Feng, F.: A new order relation on fuzzy soft sets and its application. Soft Comput. **17**(1), 63–70 (2013)

7. Herawan, T., Deris, M.M.: Soft decision making for patients suspected influenza. In: Taniar, D., Gervasi, O., Murgante, B., Pardede, E., Apduhan, B.O. (eds.) ICCSA 2010, Part III. LNCS, vol. 6018, pp. 405–418. Springer, Heidelberg (2010)

8. Hu, Y., Guo, C., Ngai, E.W.T., Liu, M., Chen, S.: A scalable intelligent non-content-based spam-filtering framework. Expert Syst. Appl. **37**(12), 8557–8565 (2010)

9. Idris, I., Selamat, A., Omatu, S.: Hybrid email spam detection model with negative selection algorithm and differential evolution. Eng. Appl. Artif. Intell. **28**, 97–110 (2014)

10. Inuiguchi, M., Yoshioka, Y., Kusunoki, Y.: Variable-precision dominance-based rough set approach and attribute reduction. Int. J. Approximate Reasoning **50**(8), 1199–1214 (2009)

11. Ali, M.I.: A note on soft sets, rough soft sets and fuzzy soft sets. Appl. Soft Comput. J. **11**(4), 3329–3332 (2011)

12. Jain, K.: A hybrid approach for spam filtering using support vector machine and artificial immune system, pp. 5–9 (2014)

13. Li, P., Wu, J., Qian, H.: Groundwater quality assessment based on rough sets attribute reduction and topsis method in a semi-arid area, china. Environ. Monit. Assess. **184**(8), 4841–4854 (2012)

14. Ma, X., Sulaiman, N., Qin, H.: Parameterization value reduction of soft sets and its algorithm. In: 2011 IEEE Colloquium on Humanities, Science and Engineering, CHUSER 2011 (Chuser), pp. 261–264 (2011)

15. Meng, D., Zhang, X., Qin, K.: Soft rough fuzzy sets and soft fuzzy rough sets. Comput. Math. Appl. **62**(12), 4635–4645 (2011)

16. Rathi, M., Pareek, V.: Spam mail detection through data mining a comparative performance analysis. Int. J. Mod. Educ. Comput. Sci. **5**, 31–39 (2013)

17. Salama, M.A., Hassanien, A.E., Revett, K.: Employment of neural network and rough set in meta-learning. Memetic Comput. **5**(3), 165–177 (2013)

18. Senan, N., Ibrahim, R., Nawi, N.M., Yanto, I.T.R., Herawan, T.: Soft set theory for feature selection of traditional malay musical instrument sounds. In: Zhu, R., Zhang, Y., Liu, B., Liu, C. (eds.) ICICA 2010. LNCS, vol. 6377, pp. 253–260. Springer, Heidelberg (2010)

19. Son, C.S., Kim, Y.N., Kim, H.S., Park, H.S., Kim, M.S.: Decision-making model for early diagnosis of congestive heart failure using rough set and decision tree approaches. J. Biomed. Inform. **45**(5), 999–1008 (2012)

20. Whissell, J.S., Clarke, C.L.: Clustering for semi-supervised spam filtering categories and subject descriptors. In: Proceedings of the 8th Annual Collaboration, Electronic messaging, Anti-Abuse and Spam Conference, pp. 125–134. ACM (2011)

21. Zhang, Z.: A rough set approach to intuitionistic fuzzy soft set based decision making. Appl. Math. Model. **36**(10), 4605–4633 (2012)

22. Zhu, D., Li, Y., Nie, H., Huang, C.: A decision-making method based on soft set under incomplete information. In: 2013 25th Chinese Control and Decision Conference (CCDC), pp. 1620–1622. IEEE (2013)

Combining Feature Selection with Decision Tree Criteria and Neural Network for Corporate Value Classification

Ratna Hidayati$^{(\boxtimes)}$, Katsutoshi Kanamori, Ling Feng,
and Hayato Ohwada

Department of Industrial Administration, Faculty of Science and Technology,
Tokyo University of Science, 2641 Yamazaki,
Noda-shi, Chiba-ken 278-8510, Japan
7415623@ed.tus.ac.jp, {katsu,ohwada}@rs.tus.ac.jp,
fengl@rs.noda.tus.ac.jp

Abstract. This study aims to classify corporate values among Japanese companies based on their corporate social responsibility (CSR) performances. Since there are many attributes in CSR, feature selection with decision tree criteria is used to select the attributes that can classify corporate values. The feature selection found that 41 % of 37 total attributes, or only 15 attributes, are needed to classify corporate values. The accuracy of building the tree used to find the 15 attributes is low. To increase the accuracy, the attributes are trained in a neural network. The accuracy of the decision tree is 0.7, and the accuracy of the neural for training the 15 attributes increased to 0.75. To sum up, this study found, companies with higher corporate values seek to enhance their CSR activities or to empower secondary stakeholders. In contrast, companies with low corporate values still focus their CSR activities on primary stakeholders.

Keywords: Corporate social responsibility · Corporate value · Feature selection · Decision tree · Neural network

1 Introduction

Making a company behave in a socially and environmentally responsible manner while striving for its economic goals is no simple task. It has become a necessity for companies to deal with issues that concern all kinds of stakeholders. Corporate social responsibility (CSR) involves all of the proper social, environmental and economic actions that a firm must incorporate to satisfy the concerns of stakeholders and the financial requirements of shareholders [1]. Therefore, CSR actions ought to be correlated with the financial state and outcomes of firms. Many studies have concentrated on the potential linkage between CSR and financial performance. To deal with the problem, one study suggested conducting research on the relationship between CSR and financial performance in different markets and regions, since there are numerous different factors that differentiate how companies operate, and how this can influence the relationship. They argue that most studies in this area have focused on the US Stock

© Springer International Publishing Switzerland 2016
H. Ohwada and K. Yoshida (Eds.): PKAW 2016, LNAI 9806, pp. 31–42, 2016.
DOI: 10.1007/978-3-319-42706-5_3

Exchange and that this limits the opportunity to generalize the results as the degree of governance, environmental policies, and business practices vary globally [2].

As one of the leading countries in CSR, the Japanese understanding CSR is linked with the country's history of industrial pollution. As a result, the area of CSR in which Japanese companies have made the most progress is the environment. The present study aims to classify corporate values calculated by the Ohlson model among Japanese companies based on many dimensions of CSR, not just the environment dimension. The Ohlson model is used to estimate income based on expected profits. Using the Ohlson model, we wanted to explore the relation of CSR performance and corporate values from a long-term perspective, since many previous studies only focused on the relation of CSR and financial performance using a market-based approach or from a short-term perspective. The present study used a decision tree as an algorithm to classify corporate values because of its several benefits, including being easy to understand and interpret. However, due to its low accuracy, the present study combines the decision tree with a neural network, a method that is recognized as having good accuracy in some areas. The predictive accuracies obtained with neural networks are often significantly higher than those obtained with other learning paradigms, particularly decision trees. However, decision trees have been preferred when having a good understanding of the decision process is essential, as in this study.

The study is organized as follows. The next section reviews previous studies on CSR and financial performance. The third section presents our data and methodology, and the fourth section discusses the results and analyses. This is followed by a concluding section and a statement regarding further research.

2   Related Works

Understanding the causal relationship between CSR and financial performance is quite difficult. Many studies have reported positive, negative or neutral impacts of CSR on financial performance. The inconsistency in results is caused by differences in methodologies, approaches, and the selection of variables [3]. In addition, few studies have based their estimations on corporate value. In fact, it is important to focus on corporate value, rather than stock returns, which have been at the center of most market-based financial studies. Many studies also examined only specific dimension of CSR. In fact, there are many indicators or dimensions of CSR performance.

For example, focusing on stakeholder management, one study found little evidence of causality between financial performance and social performance. The study concluded that strong stock market performance simply leads to greater company investment in aspects of CSR focusing on employee relations. Hence, CSR activities do not affect financial performance [4]. Another study concluded that CSR levels and their relationship with profit vary by industry. Moreover, stock market measures and accounting measures also respond differently to CSR measures, with stock market measures being better than accounting measures [5].

In Japan, CSR has become an indispensable element of corporate management. Domestically, many companies are undertaking efforts in the areas of the environment, human rights, and women's advancement. Overseas, the top area they are addressing is the environment as well [6]. Many Japanese companies put the environment high on their agendas. The environmental efforts of CSR are linked with the history of industrial pollution in Japan. Hence, many Japanese companies have reported on their environmental performance for years, such as publishing information concerning their environmental policies, mainly out of accountability considerations or as part of their risk management. The market for environmentally friendly products and services is also growing in Japan [7]. Based on this, it is clear that the strength of Japanese corporations in doing CSR activities stems from the environmental aspects.

Most companies, not only those in Japan, realize that CSR is linked to their reputation and brand identity. Using structural equation modeling (SEM), one study found that CSR positively affects customer satisfaction and loyalty. The importance order of CSR factors is as follows: consumer protection, philanthropic responsibility, legal responsibility, ethical responsibility, economic responsibility, and environmental contribution [8]. CSR has become a useful tool for gaining customer loyalty in today's markets. Customers who identify more strongly with a company tend to purchase from it more and to recommend both the company and its products more often [9]. Retaining loyal customers can certainly increase profits. It is plausible that the more loyal customers companies have, the higher their profits.

Although CSR is important for winning customer loyalty, not all companies have a CSR department in their organizations. According to [10], the size of corporate community involvement (CCI) programs is related to the allocation of responsibility in the company's organizational structure. Companies with small CCI programs manage their community involvement through their central administrative functions rather than through a marketing/HR or CSR department. The allocation of responsibility for CCI is also related to the industry in which the company operates. In Japanese companies, five types of units have been established as units related to CSR activities: a CSR department, legal and regulatory compliance department, IR department, social contribution department, and environmental department. Of these, an environmental department was confirmed to have been the earliest established. In 2005, or around ten years ago, only 25.6 % of 749 companies in Japan had established this CSR department [11]. This result indicates that having a CSR department is still considered debatable in Japan.

3 Data and Methodology

3.1 Data

CSR data was obtained from Toyo Keizai and financial data from the Nikkei. In machine learning, it is critical to input the right data. The process of getting data ready for a machine learning algorithm can be seen in the following figure:

According to the Toyo Keizai database for 2015, there are more than 100 attributes. These attributes of CSR data are grouped into three categories: (1) employment and human resources (HR), (2) CSR in general, and (3) the environment. Data on employment and HR include such information as the number of employees, re-employment system, health and safety management system, and flextime system. Data on CSR in general

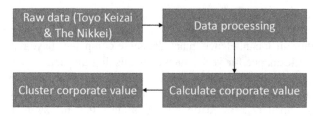

Fig. 1. Data collection

include the CSR department, CSR officers, stakeholder engagement, and ISO 26000. Data on the environment include the environmental department, climate change initiatives, environmental policy, and green purchasing.

Since this study explores Japanese companies, we did not use all attributes. We selected attributes based on the ISO 26000 framework, a CSR framework that Japan officially adopted as a national standard. In the end, we used only 37 attributes. All of these attributes are categorical. For example, regarding the environmental department attribute, companies were asked about the existence of an environmental measures department. The answers are 1 for a full-time department, 2 for concurrent department, 3 for none, 4 for other, and 0 for no answer.

Originally, we also had data on more than 1000 companies from Toyo Keizai. However, to calculate corporate values by the Ohlson model, we need financial information from the Nikkei (Nikkei Economic Electronic Databank System) for 2015. Unfortunately, not all companies provide their financial data. The corporate value is calculated by the Ohlson Model based on the income approach. The income approach is the most valid way to measure the value of a business or business interest. In addition, the Ohlson model approach is good for policy recommendations [12]. The formulation of the Ohlson model is given below:

$$V_t = b_0 + \sum_{i=1}^{\infty} \frac{E_t\left[X_{t+i}^a\right]}{(1+r)^i} \tag{1}$$

Where, V_t: corporate value, b_0: equity at book value, $E_t\left[X_{t+i}^a\right]$: residual profit, and r: cost of equity.

In the end, we obtained values for only 260 companies. The highest value is 1,766,976, and the lowest value is −51,160.2. Next we clustered the corporate values to make them easier to classify. We used simple k-means to cluster the corporate values. This is an algorithm to classify or group the objects based on attributes or features into k (a positive integer) groups. We used the elbow method to determine the optimal number for k in k-means clustering. This is the oldest method for determining the true number of clusters in a data set, albeit inelegantly. However, as a heuristic method, it may not work well in any particular case. Sometimes, there is more than one elbow, or no elbow at all. The idea of the elbow method is to choose the k at which the SSE (sum of squared error) decreases abruptly. This produces an "elbow effect" in the graph.

In this study, we first use a hierarchical method or Ward's method to define the number of clusters. We then use the k-means procedure to actually form the clusters.

Ward's is the only agglomerative clustering method that is based on a classical sum-of-squares criterion, producing groups that minimize within-group dispersion at each binary fusion. In addition. Ward's method shares the total error sum-of-squares criterion with K-means partitioning, which is widely used to directly cluster observations in Euclidean space and hence is used to partition the observation set [13].

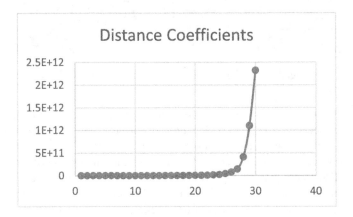

Fig. 2. Ward's method

Using this method, we found that the k in k-means clustering is 3, or 30–27, where 30 is the number of cases and 27 is the step of the elbow. We identified step 27, where the "distance coefficients" starts to make a bigger jump. The k-means provide results from 260 samples, with a low value of 205, a medium value of 39, and a high value of 16. The final data consist of 260 samples and 37 features with three classes.

3.2 Methodology

Many studies have combined a decision tree and a neural network. We can classify the literature of these hybrid systems combining decision trees and neural networks into two groups. The first group includes studies in which decision trees are used for constructing the neural network's structure. An example of this group is composing a three-layer neural network using decision trees, where the neurons of the hidden layer are determined and constructed by the decision tree [14]. The second group includes studies in which the neural networks are used right from the beginning of constructing the tree. This means that the decision tree and the neural network are working together from the beginning of the combining process. For example, using the C4.5 algorithm and at the same time applying a single-layer perceptron model to the inner node of the three. These neural networks thus have the same structure in every node, but they are individual objects in each node. Our study belongs to the first group, where a decision tree is used as input to the neural network. The process of combination is explained in the following figure.

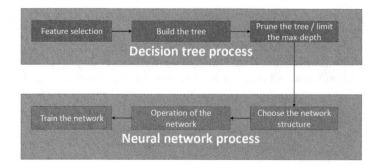

Fig. 3. Combining decision tree and neural network

There are two general processes, the decision tree process and the neural network process. In the decision tree process, features are selected using decision tree criteria. This phase aims to determine the maximum attributes that can classify corporate values among Japanese companies and serve as input to a neural network. After determining the number of attributes that can perform best when building the tree, the next question is how to obtain those attributes. In this study, we determine the attributes by building a tree to find them. Although the decision trees generated by methods such as ID3 and C4.5 are accurate and efficient, they often suffer the disadvantage of providing very large trees that make them incomprehensible to experts [15]. To solve this problem, researchers in the field have considerable interest in tree pruning. Another way to avoid overfitting is to limit the maximum depth of the tree and to make several attempts until we identify the attributes that can give the best accuracy or can clearly distinguish the classification task.

The second phase is devoted to the neural network. After we obtain the attributes from the tree, sometimes the accuracy is low, especially if we have allowed the tree to become large. Hence, the process of classification is assigned to the neural network. In the neural network, the first thing is to choose the network structure. We need to decide how many neurons will be in the input and output layers, how many hidden layers the network is going to have, and how many neurons will be in each of these hidden layers. The next process is to specify the operation of the network, starting with the learning phase. The time required for the learning phase depends on the size of the neural network, the number of patterns to be learned, and so on. After we deal with all of this, we can train our neural network. In this study, the network structure of the neural network is determined based on the decision tree results (Figs. 1 and 2).

4   Results

4.1   Feature Selection with Decision Tree Criteria

As seen in Fig. 3, the process of combining a decision tree and a neural network starts with feature selection. The result of feature selection can be seen in the following figure:

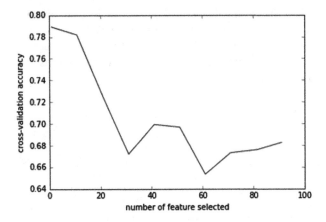

Fig. 4. Feature selection with decision tree criteria

According to Fig. 4, the optimal number of features is only 1 % of the total. This means that to classify corporate values based on CSR performance, the highest accuracy is achieved if we only use 1 % of the total features or essentially a single attribute. However, we don't want to use a single attribute, we want to use as many attributes as possible to classify the corporate value. In this study, we will use 41 % of the total attributes, representing around 15 attributes.

In Fig. 4, we only know the number of attributes and the given accuracy. To determine the attributes, we need to build the tree. The problem with predicting the 15 attributes is that if we allow the tree to become bigger, we will need to deal with overfitting in one of two ways. In this study we do not need to prune the tree, but with the number of maximum depths at 8, the total attributes in the tree number 15, or 41 % of the total number of attributes.

Table 1. Feature selection and several maximum-depth results

Feature Selection (in cross-validation)		Limit the max-depth (in training set)		
percentile (%)	accuracy	number of max-depth	accuracy	number of features
1	0.789	1	0.788	1
11	0.782	2	0.788	3
21	0.725	3	0.788	6
31	0.672	4	0.804	8
41	0.699	5	0.819	10
51	0.697	6	0.838	13
61	0.653	7	0.854	14
71	0.673	8	0.854	15
81	0.676	9	0.854	16
91	0.682	10	0.854	14

Based on Table 1, increasing the number of features does not improve the accuracy. We can see from the feature selection table that above 41 %, the accuracy goes down. Increasing the number of maximum depths does not improve the accuracy either. The maximum depth remains stable at above 7. The following figure is a tree based on a maximum depth of 8.

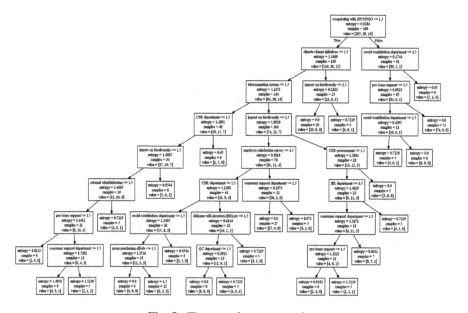

Fig. 5. The tree of corporate value

Using max-depth 8, we found 15 attributes from the tree. The attributes are listed in the following table. These attributes will be an input for the neural network (Table 2).

Some attributes in the tree above are repeated. This indicates that 15 attributes are sufficient to classify corporate value among Japanese companies. Adding more attributes does not improve the accuracy of classification. Although some studies have already proven that CSR activities can influence financial performance, the present study tried to investigate further using corporate value. Based on Fig. 5, the following points can be concluded:

1. Almost all companies with high or medium value are cooperating with NPOs/NGOs, conducting employee satisfaction surveys, and considering external whistleblowing, and have CSR procurement and HR departments for managing the diversity of their human resources.
2. In contrast, not all companies with low corporate value consider these CSR factors. However, almost all of them have social-contribution and consumer-support departments.

Table 2. Name of attributes

No.	Name of attributes
1	cooperating with NPO/NGO
2	climate change initiatives
3	social contribution department
4	telecommuting system
5	impact on biodiversity
6	pro bono support
7	CSR department
8	employee satisfaction survey
9	CSR procurement
10	external whistleblowing
11	consumer support department
12	HR department
13	dialogue with investors, ESG, etc.
14	green purchasing efforts
15	QC department

3. In addition, compared with companies with low corporate value, companies that have medium or high corporate value start to implement a telecommuting system and commit to pro bono.
4. Not all companies with low, medium or high corporate value based on the tree have a CSR department. In the context of environmental issues, all of them are considering the impact on biodiversity of business activities, green purchasing efforts, and climate change initiatives.

The remaining attributes, i.e. a QC department, dialogue with investors, ESG, and so on, are only able to distinguish among companies with low corporate value.

4.2   Cross-Validation

In this section, we evaluate the model by cross-validation. Cross-validation is a technique to evaluate predictive models by partitioning the original sample into a training set to train the model and a test set to evaluate it. K-fold cross validation is a common technique for estimating the performance of classifier. This study uses 10-fold cross validation. The result for each classifier can be seen in the following table.

Based on Table 3, both the decision tree and the neural network are generally able to give a good result. To make it relatively simple to understand how a decision tree and neural network classify corporate values, we will look at the confusion matrix result in Table 4.

A confusion matrix is a table that is often used to describe the performance of a classification model or classifier on a set of test data for which the true values are known. Table 4 indicates the performance of the neural network (NN1) in training all

Table 3. Classification report

Class	Precision			Recall			F1 score			Support
	DT	NN1	NN2	DT	NN1	NN2	DT	NN1	NN2	
0	0.82	0.84	0.84	0.81	0.86	0.88	0.82	0.85	0.86	250
1	0.32	0.35	0.36	0.33	0.38	0.33	0.32	0.37	0.34	39
2	0.13	0.1	0.14	0.12	0.06	0.06	0.13	0.08	0.09	16
Avg./Total	0.7	0.72	0.72	0.7	0.73	0.75	0.7	0.73	0.73	260

Table 4. Confusion matrix

	decision tree accuracy: 0.7			NN1 accuracy: 0.73			NN2 accuracy: 0.75		
class/classifier	0	1	2	0	1	2	0	1	2
0	168	27	10	176	24	5	182	18	5
1	18	14	7	20	15	4	25	13	1
2	10	3	3	12	3	1	10	5	1

attributes is also better than the decision tree. Moreover, using a neural network (NN2) after feature selection based on decision tree criteria can give the best accuracy.

5 Conclusion and Further Research

Increasing pressure to continuously manage their social responsibilities has forced companies to dedicate more time and resources to CSR activities in order to meet stakeholder expectations [16]. Based on this, many studies have tried to determine the relationship between CSR activities and financial performance. An early study that examined the relationship between a company's CSR activities and its financial performance concluded that firms low in social responsibility experienced weaker financial performance while noting that low CSR performance also could expose such companies to risks to a larger extent than high-performing firms [17].

This study contributes to the existing literature on the relationship between CSR and financial performance in several ways. First, our results showed that companies with high and medium corporate value in doing CSR activities seek to further empower people and societies or to enhance their CSR activities. This is achieved by cooperating with NPO/NGOs and committing to pro bono. Pro bono is mostly used to describe professional work delivered voluntarily and without payment or at a highly reduced fee. These two attributes can clearly distinguish these companies from those that have low corporate value. In addition, companies with high and medium corporate value try to promote work-life balance by implementing telecommuting system. Second, companies with low value still focus on doing CSR activities for their internal resources or performing direct actions for their stakeholders, such as dialogue with investors, ESG, etc.

Finally, we can generally conclude that many Japanese companies are considering environmental issues, and that no matter what their value is their efforts are almost the same. Many Japanese companies do not have a CSR department to manage their social activities. This result is consistent with a previous study stating that not all companies handle their social activities through a CSR department. They can manage their social activities through the Marketing/HR department or as an administration function.

For further study, we would like to implement a second group method in which neural networks are used right from the beginning of constructing the tree. This means that the decision tree and the neural network are working together from the beginning of the combination process. Using a decision tree that is easy to understand is sufficient to create an appropriate model to investigate the characteristics of Japanese companies doing CSR activities based on their corporate values. However, we need to improve the ability of the model to perform the necessary classification. On the other hand, neural networks are "black boxes" methods. It is hard to look at into the network and exactly figure out what it has learnt. However, neural networks are better at accuracy. By combining decision tree and neural network the process of learning is easy to understand and the accuracy obtained is also good. This is useful method when the process of learning is more important than the classification accuracy.

References

1. Ingley, C., Mueller, J., Cocks, G.: The financial crisis, investor activists and corporate strategy: will this mean shareholders in the boardroom? J. Manag. Gov. **15**, 557–587 (2011). doi:10.1007/s10997-010-9130-9
2. Peng, C.W., Yang, M.L.: The effect of corporate social performance on financial performance: the moderating effect of ownership concentration. J. Bus. Ethics **123**(1), 171–182 (2014)
3. Girerd-Potin, I., Jimenez-Graces, S., Louvent, P.: Which dimensions of social responsibility concern financial investors? J. Bus. Ethics Univ. Grenoble Aples **10**(1), 559–576 (2013)
4. Nelling, E., Elizabeth, A.E.: Corporate social responsibility and financial performance: the "virtuous circle" revisited. Rev. Quant. Finan. Acc. **32**, 197–209 (2009). doi:10.1007/s11156-008-0090-y. Web Published online: 14 May 2008. Springer Science + Business Media, LLC (2008)
5. Beliveau, B., Cottril, M., O'Neil, H.M.: Predicting corporate social responsiveness: a model drawn from three perspectives. J. Bus. Ethics **13**, 731–738 (1994). Kluwer Academic Publishers. Printed in the Netherlands
6. Zentaro, K., Taku, H.: Issues and Prospects for CSR in Japan Analysis of Japan's CSR Corporate Survey (2015). http://www.tokyofoundation.org/en/articles/2015/issues-and-prospects-for-csr. Accessed 18 January 2016
7. Yamada, S.: Environmental measures in Japaneses enterprises: a study from an aspect of socialisation for employees. In: Szell, G., Tominaga, K. (eds.): The Environmental Challenges for Japan and Germany: Intercultural and Interdisciplinary Perspectives, pp. 297–322. Peter Lang, Frankfurt/Main (2004)
8. Chung, K.H., Yu, J.E., Choi, M.G., Shin, J.I.: The effects of CSR on customer satisfaction and loyalty in China: the moderating role of corporate image. J. Econ. Bus. Manag. **3**(5), 542–547 (2015). doi:10.7763/JOEBM.2015.V3.243

9. Ahearne, M., Bhattacharya, C.B., Gruen, T.: Antecedents and consequences of customer – company identification: expanding the role of relationship marketing. J. Appl. Psychol. **90**(3), 574–585 (2010)
10. Brammer, S., Millington, A.: The effect of stakeholder preferences, organizational structure and industry type on corporate community involvement. J. Bus. Ethics **45**, 213–226 (2003). Kluwer Academic Publishers. Printed in the Netherlands
11. Koh, Y.: CSR at Japanese companies as seen in changes in administrative departments. J. Econ. Bus. Manag. **3**(11), 1054–1060 (2015). doi:10.7763/JOEBM.2015.V3.333
12. Hand, J.R.M., Landsman, W.R.: Testing the Ohlson Model: v or not v, That is the Question. Working Paper, University of North Carolina, Chapel Hill (1999)
13. Murtagh, F., Legendre, P.: Ward's hierarchical agglomerative clustering method: which algorithms implement ward's criterion? J. Classif. **31**, 274–295 (2014). doi:10.1007/s00357-014-9161-z
14. Sethi, I.K.: Entropy nets: from decision trees to neural networks. Proc. IEEE **78**, 1605–1613 (1990). doi:10.1109/5.58346
15. Quinlan, P.F.: Simplifying decision trees. Int. J. Hum. Comput. Stud. **51**, 497–510 (1999)
16. McGuire, J.B., Sundgren, A., Schneeweis, T.: Corporate social responsibility and firm financial performance. Acad. Manag. J. **31**(4), 854–872 (1988)
17. González-Rodríguez, R., Carmen Díaz-Fernández, M., Simonetti, B.: The social, economic and environmental dimensions of corporate social responsibility: the role played by consumers and potential entrepreneurs. Int. Bus. Rev. **1**(1), 1–25 (2015)

Learning Under Data Shift for Domain Adaptation: A Model-Based Co-clustering Transfer Learning Solution

Santosh Kumar, Xiaoying Gao[(⊠)], and Ian Welch

School of Engineering and Computer Science, Victoria University of Wellington,
Wellington, New Zealand
{santosh.kumar,xiaoying.gao,ian.welch}@ecs.vuw.ac.nz

Abstract. Data shifting in machine learning problems violates the common assumption that the training and testing samples should be drawn from the same distribution. Most of the algorithms which provide the solution for data shifting problems first try to evaluate the distributions and then reweight samples based on their distributions. Due to the difficulty of evaluating a precise distribution, conventional methods cannot achieve good classification performance. In this paper, we introduce two types of data-shift problems and propose a model-based co-clustering transfer learning based solution which consistently deals with both scenarios of data shift. Experimental results demonstrate that our proposed method achieves better generalization and running efficiency compared to traditional methods under data or covariate shift setting.

1 Introduction

The methods based on machine learning techniques usually face a major challenge when applied in the wild: The circumstances under which the method was developed will differ from those in which we use the method. An example could be a complex email spam filtering method that needed a few years to develop. Will this method be useful, or will it require to be modified because the types of spam have developed since the system was original built? Apparently any real world data analysis is bothered with such difficulties, which occur for purposes extending from the bias presented by preliminary design, to the poor irreproducibility of the testing circumstances at training period. In an ideal form, any of these problems can be recognized as cases of dataset shift, wherever the joint distribution of inputs and outputs varies between training and testing stage [1].

In the real world, the circumstances in which we use the methods will change from the circumstances in which they were generated. Typically conditions are not stationary, and sometimes the challenges of meeting the development situations to the use situations are too large or too costly [2]. More formally, giving some data, and some modelling framework, a model can be determined. This model can be used for performing predictions $P(y/x)$ for some targets y given some distinct x. Though, if there is a chance that something may have varied between training and testing conditions, it is necessary to examine if a different predictive model should be used. For this, it is

H. Ohwada and K. Yoshida (Eds.): PKAW 2016, LNAI 9806, pp. 43–54, 2016.
DOI: 10.1007/978-3-319-42706-5_4

important to improve a perception of the fitness of appropriate models in the circumstance of such variations. Understanding of how robust to model the basic changes will enable the better description of the outcome of these differences. There is the problem of what requires being done to implement the resulting method. Does the training process itself require to be replaced, or is there simply post hoc processing that can be done to the learned pattern to estimate the change?

In this paper, we explain domain adaptation problem for data classification by learning under data/covariate shift. This solution comprises a novel model-based co-clustering transfer learning approach. Below, we give two types of classification scenarios relating to data/covariate shift. They are what we concentrate on in this paper.

- Scenario 1. Classification problems include training examples, testing examples, and auxiliary examples, where training examples and testing examples are formed from the same distribution whereas the auxiliary examples are formed from a different distribution. Also, the training set size is small.

  In the practical situations, many of classification problems refer to Scenario 1. For example, suppose we want to create a Web-page classification model. The Web data employed in training a Web-page classification model can be quickly outdated when employed to the Web someday later as the topics on the Web change regularly. Usually, new data are valuable to label and therefore, their quantities are restricted due to cost concerns. How to precisely classify the new test data by obtaining the most use of the old data becomes a significant problem.

- Scenario 2. Classification problems include training examples and testing examples, where training examples and testing examples are formed from different distributions. There are no auxiliary examples.

  For example, suppose we are applying a learning method to produce a model that predicts the side effects of medication for a given patient. Because the medication is not given randomly to individuals in the normal population, the possible training examples are not a random example from the population. Consequently, the training examples and testing examples are formed from distinct distributions. How to precisely classify the testing data by applying the training examples becomes a critical problem.

In this paper, we address the two types of domain-based data shift problems by training on a newly constructed dataset following approximately the target distribution and implementing a model based co-clustering transfer learning solution for data shift problem.

2   Related Concepts and Related Work

2.1   Related Concepts

In this section, we present some symbols and descriptions that are used in this paper. First of all, we give the explanations of "domain", "task," and "data shift" respectively [3].

Definition 1 (domain). In this paper, a domain D consists of two components: a feature space F and a marginal probability distribution $P(X)$, where $X = \{x_1, \ldots, x_n\} \in F$.

Definition 2 (task). Given a specific domain $D = \{F, P(X)\}$, a task contains two components: a label space Y and an objective function $f(.)$ (denoted by $\wp = \{Y, f(.)\}$), which is not observed but can be learned from the training data, which consist of pairs $\{x_i, y_i\}$, where $x_i \in X$ and $y_i \in Y$. The function $f(.)$ can be used to predict the corresponding label, $f(x)$ of a new instance x. From a probabilistic viewpoint, $f(x)$ can be written as $P(y \mid x)$. In this section, we denote the source domain data by $D_s = \{(x_{S_1}, y_{S_1}), \ldots, (x_{S_{ns}}, y_{S_{ns}})\}$, where $x_{S_i} \in X_S$ is the data instance and $y_{S_i} \in Y_S$ is the corresponding class label. Similarly, we denote the target domain data by $D_T = \{(x_{T_1}, y_{T_1}), \ldots, (x_{T_{nT}}, y_{T_{nT}})\}$, where the input x_{Ti} is in X_T and $y_{T_i} \in Y_T$ is the corresponding output. In most cases, $0 \le n_T \ll n_S$.

Definition 3 (Data shift). Covariate shift refers to the learning settings that have the following features: (1) source domain and target domain have the same feature and label spaces; that is $X_S = X_T$ and $Y_S = Y_T$. (2) Source domain and target domain have different feature distribution; that is, $P_S(X) \ne P_T(X)$. (3) Source domain and target domain have the same concept; that is $f_s(.) = f_t(.)$ or $P_S(Y \mid X) = P_T(Y \mid X)$ [4].

It is worthwhile to note that there can be multiple auxiliary data sets in classification problems under data shift and their feature distributions can be different. In addition, from the definition of covariate shift, we can see that the two scenarios described in Sect. 1 do belong to covariate shift, because, for Scenario 2, testing samples can be considered as target samples and training samples can be considered as auxiliary samples.

2.2 Dataset Shift and Transfer Learning

Dataset shift and transfer learning are closely related. Transfer learning examines the problem of how knowledge can be taken from some only partially related training scenarios and used to present better prediction in one of those scenarios than would be obtained from that scenario alone [5]. Therefore, dataset shift consists of the case where there are only two scenarios, and one of those scenarios has no training targets. Multi-task learning is also related. In multitask learning, the acknowledgement for a provided input on a type of tasks is achieved, and knowledge between tasks is applied to support the prediction. Multitask learning can be considered of a particular case of transfer learning where there is some commonality in training covariates between tasks, and where the covariates have the same meaning over scenarios (hence domain shift is prevented).

2.3 Related Work

As explained before, covariate shift includes the two scenarios outlined above. In Scenario 1, auxiliary examples are utilised to enhance the performance of classifiers. Previously, Wu and Dieterich [6] introduced an image classification algorithm using both small training data and plenty of auxiliary data. They illustrated some improvement by using the auxiliary data. However, they did not give a quantitative study using various auxiliary examples. Liao et al. [7] improved learning with auxiliary data using active learning. Rosenstein et al. [8] suggested a hierarchical naïve Bayes method for transfer learning using auxiliary data and considered when transfer learning would advance or reduce the performance. Dai et al. [9] recommended a covariate shift-related algorithm, TrAdaBoost, which is an expansion of the AdaBoost algorithm, to address the inductive transfer learning problems. TrAdaBoost implies that the source and target domain data use the same set of features and labels, but the distributions of the data samples in the two domains are distinct. Besides, TrAdaBoost believes that due to the difference in distributions between the source and the target domains, some source domain data may be beneficial in learning for the target domain but some of them may not and could even be damaging. It tries to reweight the source domain data iteratively to overcome the effect of the "bad" source data although supporting the "good" source data to provide more to the target domain. For each round of repetition, TrAdaBoost trains the base classifier on the weighted source and target data. The error is only measured on the target data. Moreover, TrAdaBoost uses the same approach as AdaBoost [10] to update the incorrectly classified examples in the target domain while using a distinct procedure from AdaBoost to update the incorrectly classified source examples in the source domain. However, TrAdaBoost cannot deal with the case where multiple auxiliary data sets are coming from different distributions.

Concerning Scenario 2, unlabeled testing samples are utilized to enhance the performance of classifiers. Unlike semi-supervising learning problem [11], for Scenario 2, the unlabeled testing examples are under a distinct distribution from the training samples and are used to correct the sample selection bias. In previous work, most strategies intend to estimate the importance $P(x_{S_i})/P(x_{T_i})$. If we can estimate the importance for each instance, we can solve the learning problems under covariate shift. There exist various ways to estimate $P(x_{S_i})/P(x_{T_i})$.

Zadrozny [12] proposed to estimate the terms $P(x_{S_i})$ and $P(x_{T_i})$ and independently by building simple classification problems and then to determine the importance by taking the ratio of the estimated frequencies. However, estimating frequencies are known to be a difficult problem, individually in high dimensional cases. Therefore, this approach may not be effective. Huang et al. [13] proposed a kernel-mean matching (KMM) algorithm to learn directly by coordinating the means between the source domain data and the target domain data in a reproducing kernel Hilbert space (RKHS). KMM is shown to work well if tuning parameters such as the kernel width are determined appropriately. Thus, the importance estimation problem is now relocated to the model selection problem. Standard model selection methods such as cross-validation, nevertheless, are gradually biased under covariate shift. Therefore, KMM cannot be directly applied in the cross-validation [14] framework. Unlike KMM, Sugiyama et al. [15] introduced an algorithm known as Kullback-Leibler importance estimating procedure (KLIEP), which is implemented with a simple model selection method. KLIEP can be integrated with cross-validation to perform model selection automatically in two steps: (1) determining the weights of the source domain data; (2) training models on the reweighted data.

In this paper, we introduce a novel method called MIDS to deal with classification problems under covariate shift. The formal definition of MIDS and its formation method will be presented in the next section. Unlike previous transfer learning methods, our method can consistently deal with both scenarios and the circumstances where multiple auxiliary data sets are coming from different distributions. Moreover, unlike the above sample reweighting techniques, we do not consider distributions but match sample numbers between target set and auxiliary set in each feature subspace; we do not re-weight examples but build a new training set following approximately the target distribution.

3 Problem Formulation

In this section, we use two data sets, target set and auxiliary set. Our goal is to design a method that can build a co-clustering transfer learning model from auxiliary examples according to target distribution.

3.1 Basic Ideas of the Model-Based Co-clustering Method

In model-based co-clustering, it is assumed that the data arises from a mixture of underlying probability distributions, where each component k of the mixture represents a cluster. Hence the matrix data $x = (x_1, \ldots, x_n)$ is assumed to be independent and identically distributed $(i.i.d)$ and generated from a probability distribution with the density

$$f(x_i; \theta) = \sum_k \pi_k f_k(x_i; \alpha) \tag{1}$$

Where f_k is the density function of the kth component. Generally, these densities belong to the same parametric family (Gaussian). The parameters $(\pi_k)_{k=1}^g$ give the probabilities that an observation belongs to the kth component and the number of components in the mixture is g which is assumed to be known. The parameter θ of this model is the vector $(\pi_1, \ldots, \pi_g, \alpha)$ containing the parameter α and the mixing proportions $(\pi_1, \ldots, \pi_g)$. Using this, the density of the observed data x can be expressed as

$$f(x_i; \theta) = \prod_i \sum_k \pi_k f_k(x_i; \alpha) \tag{2}$$

3.2 Model-Based Co-clustering Transfer Learning Method for Data Shift Problem

We present a brief description of Model-Based Co-clustering transfer learning following the same notation in [16]. Given a set of n data points, we represent them by a $n * d$ matrix X, such that X_i denotes the ith data point and X_i^j represents its jth feature value. In the model co-clustering, every point X_i has a corresponding $k-$ dimensional boolean membership vector V_i where k is the desired number of clusters. The hth component V_i^h of this membership vector is a binary value indicating whether X_i belongs to the hth

cluster. So, multiple 1's encode that the point belongs to several clusters. In this algorithm, the probability of generating all data points is

$$p(X \mid \Theta) = p(X \mid V, A) = \Pi_{i,j} p\left(X_i^j \mid V_i, A^j\right) \tag{3}$$

Where A is the so-called activity matrix of this model. In the model based co-clustering, every element A_h^j is interpreted as the activity of cluster h while generating the jth feature of data. In this model, $\Theta = \{V, A\}$ are the parameters of p and X_i^j's are conditionally independent given V_i and A_j. Furthermore, it is assumed that p can be the density function of any regular exponential family distribution, and also assume that the expectation parameter corresponding to X_i is of the form $V_i A$, so that $E[X_i] = V_i A$.

Our proposed method for model-based co-clustering transfer learning is motivated from the co-clustering method. We examine two boolean membership vectors, one for each data point and one for each feature. We describe the activity matrix in the model based co-clustering model in a diverse way that would help us to continue the idea to the co-clustering case.

We see each element of activity matrix as representing the extent that feature would participate in producing that feature if the corresponding data point belonged to the only cluster. In other words, we consider each data value as produced by partial contributions based on different clusters that the relevant data point belongs to. The membership vector simply shows which clusters take part in generating a particular feature. To continue the idea to the co-clustering problem, we assume that the value of these partial contributions (partial contribution of the feature in a data point that belongs to data cluster) is defined by which classes the corresponding feature goes to. In other words, each one of the categories that the feature belongs to have a share in the feature contribution. We consider a membership vector for each feature showing the feature clusters that this feature relates to. We also consider matrix where intimates the activity of feature cluster while generating a data point that relates to the data cluster. Using a similar notation as model based co-clustering, we would have

$$E\left[A_i^T\right] = N_i C \text{ or } E\left[A^j\right] = CN^j \tag{4}$$

Based on either of these interpretations, the probability of generating all the data points will be

$$p(X \mid \Theta) = p(X \mid V, C, N) = \prod_{i,j} p\left(X_i^j \mid V_i, C, N^j\right) \tag{5}$$

In this model, $\Theta = \{V, C, N\}$ are the parameters of p and X_i^j are conditionally independent given V_i, C and N^j. Probability density p can be any regular exponential family distribution, where the expectation parameter corresponding to X_i^j is of the form $V_i C N^j$.

3.3 Algorithms and Analysis

In this section, we propose and analyse algorithms for estimating the overlapping co-clustering model given an observation matrix X. In particular, from a given observation matrix X, we want to estimate the prior matrices α and β, the membership matrices V and N and finally the activity matrix C so as to maximize the log-likelihood of the observation matrix X, assuming $p(X, V, N, C)$, the joint distribution of (X, V, N, C).

The key idea behind the estimation is similar to most algorithms for co-clustering. Each optimization step consists of two similar sub-steps. In each sub-step, we assume that clustering for one dimension of X is fixed and we optimize the objective function considering the other dimension. In the first sub-step, we assume we have fixed column clusters and we try to find optimal row clustering which optimizes the objective function. In the second sub-step, we assume row clusters being fixed and we do column clustering.

Using this approach, each of these sub-steps would be reduced to an optimization problem similar to [17] with the difference that in each sub-step, we use information obtained from the other sub-step to enhance the clustering result. The outline of the entire optimization process is showed in Algorithm 1. Basically we use a minimization technique that alternates between updating α and β, V and N and C. Because the two sub-steps mentioned are similar, we will focus on the first sub-step: having fixed column clusters and trying to find a row clustering which optimizes the objective function. The following is the algorithm for model based co-clustering transfer learning:

While   Row or Column Clusters changes do
    {Assume having fixed column clustering}
        1.   Update α
        2.   Update V { N and C is assumed known and we try to optimize V }
        3.   Update C { V and N is assumed known and we try to optimize C }
    {Assume having fixed row clustering}
        1.   Update β
        2.   Update N { V and C is assumed known and we try to optimize N }
        3.   Update C { V and N is assumed known and we try to optimize C }
End While

---

4 Experiments

In this section, we perform experiments to test the performance of the proposed model-based co-clustering transfer learning algorithm under the data shift condition.

4.1 Datasets

Movielens dataset (movie recommendations): This is a publicly available dataset from the movie recommendation system developed at the University of Minnesota [18]. The dataset contains 100, 000 ratings for 1682 movies by 943 users. It has user ratings for every movie in the collection: users give ratings on a scale of 1–5, with 1 indicating extreme dislike and 5 indicating strong approval. There are 943 users in this dataset, and the mean and the median number of users voting on any movie are 59 and 27 respectively. Each user has rated at least 20 movies. As a result, if each movie in this dataset is represented as a vector of ratings over all the users, the vector is high-dimensional but typically very sparse. Each movie has information about the different genres that this movie belongs to. If each genre is considered as separate category or cluster, then this dataset also has naturally overlapping clusters. Many movies belong to multiple genres, e.g., Aliens belongs to 3 genre categories: action, horror and science fiction. Like in [19], we created two subsets from the Movielens dataset:

- **Mv1:** 679 movies from the 3 genres. Animation, Children's and Comedy;
- **Mv2:** 232 movies from the 3 genres. Thriller, action and adventure.

We clustered the movies based on the user recommendations to rediscover genres, based on the belief that similarity in recommendation profiles of movies gives an indication about whether they are in related genres.

Text Dataset: Reuters 21578 is currently a very widely used test collection for text categorization research. The data was originally collected and labelled by Carnegie Group, Inc. and Reuters, [19]. We created a subset from this dataset in the following way: in order to have overlapping classes in the dataset, we removed all those topics which had less than 100 documents and therefore we also removed all those documents that ended up without a topic. We then had 3149 documents belonging to at least one of the six remaining topics. We used document frequency as a feature selection criterion for decreasing the dimensionality of data to 1000. We removed words which had very high document frequencies until we got dimensionality of 1000. We call this subset TX.

4.2 The Experiment on Scenario 1

Experimental Data Construction: This experiment is performed on two data sets: movie recommendation data, and text documents, from which the target training set, the target testing set, and the auxiliary training set are constructed by the following conditions.

Condition 1. The target training set and the target testing set should follow the same distribution.

Condition 2. The auxiliary training set and the target set should follow different distributions.

Condition 3. The size of the target training set is far smaller than that of the auxiliary training set.

4.3 The Experiment on Scenario 2

Experimental Data Construction: Unlike the first scenario, in this experiment, we only construct the target training set and the target testing set. We should guarantee that the target training set and the target testing set follow different distributions. Like Scenario 1, we also use a deliberately biased procedure to construct the experimental data.

5 Results

We select auxiliary samples using a deliberately biased procedure. To describe our biased selection scheme, we need to define an additional random variable si for each point in the pool of possible training samples, where $X_i = 1$ means the ith sample is included and $X_i = 0$ indicates an excluded sample. In this paper, we discuss the classification problems under covariate shift, so we only consider the situation $p(X_i \mid V_i, N^j) = p(X_i \mid V_i)$. Below, we present the detailed method of experimental data construction. We use subsets of original datasets with the characteristic that the points in the subset have natural overlapping grouping as explained later on. Using the full data sets is computationally expensive. Therefore, we use these subsets of data in order to make the datasets computationally reasonable to run experiments. But it should be noted that using smaller datasets doesn't make the task much easier, since clustering a small number of points in a high-dimensional space is still a difficult task.

Experimental Methods. In the following, we compare our method against three other transfer learning based methods for data shift problems: co-clustering transfer learning model [16], K-Means based transfer learning method [17] and information-theoretic co-clustering (ITCC) algorithm [20]. In order to compare clustering results, we use precision, recall, and F-measure calculated over pairs of points. For each pair of points that share at least one cluster in the overlapping clustering results, these measures try to estimate whether the prediction of this pair as being in the same cluster was correct with respect to the underlying true categories in the data. Precision is calculated as the fraction of pairs correctly put in the same cluster, recall is the fraction of actual pairs that were identified, and F-measure is the harmonic mean of precision and recall.

Result Analysis. Table 1 presents the results of propose Model-based Co-clustering transfer learning method versus Co-clustering transfer learning, K-Means transfer learning and ITCC algorithms for data shift in terms of precision and recall for the datasets described in Sect. 4.1 for selected experiments. Each reported result is an average over ten trials.

Table 2 presents the same results in terms of F-Measure. Table 1 shows that for all domains, even though there is no considerable difference between these methods in terms of precision in most cases, the Model-based Co-clustering transfer learning method has the best recall by a large margin compared to the other three methods: therefore Model-based Co-clustering transfer learning method consistently outperforms the other two methods in terms of overall F-measure as shown in Table 2.

The performance of the four examined methods in terms of recall, precision and F-Measure changes as the number of movie clusters increases for Mv1 dataset. With increasing number of movie clusters, the precision performance of all four methods increases slowly but the recall performance decreases more significantly. The proposed method consistently performs better than the other three methods in terms of recall and F-Measure. It is worth noting that for any methods that generate a small number of clusters, having a good recall is not surprising. Any algorithm looking for a few clusters could have relatively good recall performance. The effectiveness of an algorithm shows itself when the number of clusters is high and the algorithm still has a good recall without significant drop in precision. Results show that for the proposed algorithm, increasing the number of movie clusters has less impact on recall compared to the other three methods. As the number of movie clusters increases, the gap in performance between proposed and the other methods in terms of recall and F-Measure grows.

Table 1. Comparison results of proposed, co-clustering and K-Means algorithm on all datasets in terms of precision and recall. Mv1 and Mv2 are two datasets described in Sect. 4.1, E1, E2, E3 and E4 represents different experiments corresponding to computed number of row and column clusters equal to (6, 6), (10, 10), (6, 10) and (6, 12).

Datasets	Precision			
	Proposed	**Co-Clustering**	**K-Mean**	**ITCC**
Mv1-E1	0.60±0.01	0.61±0.01	0.60±0.00	**0.63**±0.01
Mv1-E2	0.62±0.01	0.61±0.01	0.60±0.00	**0.65**±0.01
Mv2-E1	0.46±0.01	0.47±0.02	0.48±0.01	**0.54**±0.01
Mv2-E1	0.48±0.01	0.49±0.02	0.49±0.01	**0.57**±0.02
TX-E3	0.22±0.01	**0.23**±0.01	0.22±0.01	**0.23**±0.00
TX-E4	0.23±0.00	**0.23**±0.02	0.23±0.00	**0.23**±0.00

Datasets	Recall			
	Proposed	**Co-Clustering**	**K-Mean**	**ITCC**
Mv1-E1	**0.67**±0.02	0.59±0.09	0.48±0.02	0.19±0.01
Mv1-E2	**0.65**±0.02	0.44±0.05	0.40±0.03	0.13±0.01
Mv2-E1	**0.62**±0.02	0.51±0.07	0.39±0.01	0.23±0.02
Mv2-E1	**0.58**±0.04	0.40±0.11	0.29±0.03	0.16±0.02
TX-E3	**0.51**±0.03	0.42±0.08	0.48±0.02	0.18±0.01
TX-E4	**0.48**±0.03	0.33±0.03	0.30±0.03	0.09±0.00

Table 2. Comparison results of proposed, co-clustering and KMeans algorithm on all datasets in terms of F-Measure. *Mv1* and *Mv2* are two datasets described in Sect. 4.1, E1, E2, E3 and E4 represents different experiments corresponding to computed number of row and column clusters equal to (6, 6), (10, 10), (6, 10) and (6, 12).

Datasets	Proposed Algo.	Co-Clustering	K-Mean	ITCC
Mv1-E1	**0.65** ± 0.01	0.60 ± 0.04	0.54 ± 0.01	0.29 ± 0.01
Mv1-E2	**0.64** ± 0.01	0.51 ± 0.03	0.48 ± 0.02	0.22 ± 0.01
Mv2-E1	**0.58** ± 0.01	0.49 ± 0.04	0.43 ± 0.01	0.32 ± 0.02
Mv2-E1	**0.55** ± 0.02	0.43 ± 0.06	0.36 ± 0.02	0.25 ± 0.02
TX-E3	**0.37** ± 0.01	0.29 ± 0.02	0.31 ± 0.01	0.20 ± 0.01
TX-E4	**0.38** ± 0.01	0.27 ± 0.02	0.26 ± 0.01	0.13 ± 0.00

The performance of our proposed method improves as the number of user clusters increases, as seen in the curves for two different choices of number of user clusters, 2 and 10, as well as intermediate values not shown. As the number of user clusters increases, the performance of proposed method increases. Furthermore, we see that the performance of proposed method for all choices of number of user clusters is better than the other three methods.

Amongst the four methods, ITCC has slightly better precision in most cases but due to its poor recall, its overall performance in terms of F-Measure is significantly worse than the other methods. Results for TX dataset were similar to movie dataset results. Results show that for the proposed method, increasing the number of document clusters has less impact on recall compared to the other three methods and the gap in performance between proposed method and the other methods in terms of recall and F-Measure grows. Results also show that the performance of proposed method improves as the number of word clusters increases.

6 Conclusion

In this paper, we first proposed a model-based co-clustering transfer learning method by useing a generic alternating minimization algorithm for fitting this model to empirical data, and then we performed classification under covariate shift by training on a new data set. Our basic idea is to train a model on a newly constructed data set following approximately the target distribution. We have presented promising experimental results on real news abstracts and movie data. In particular, we have shown evidence that our proposed transfer learning method produces more accurate classification results than the existing models.

References

1. Rebbapragada, U., Bue, B., Wozniak, P.R.: Time-domain surveys and data shift: case study at the intermediate palomar transient factory. In: American Astronomical Society Meeting Abstracts, vol. 225 (2015)

2. Sajobi, T.T., et al.: Identifying reprioritization response shift in a stroke caregiver population: a comparison of missing data methods. Qual. Life Res. **24**(3), 529–540 (2015)
3. Pan, S.J., Yang, Q.: A survey on transfer learning. IEEE Trans. Knowl. Data Eng. **22**(10), 1345–1359 (2010)
4. Ioffe, S., Szegedy, C.: Batch normalization: accelerating deep network training by reducing internal covariate shift. arXiv preprint arXiv:1502.03167 (2015)
5. Quionero-Candela, J., et al.: Dataset Shift in Machine Learning. The MIT Press, Cambridge (2009)
6. Shimodaira, H.: Improving predictive inference under covariate shift by weighting the log-likelihood function. J. stat. plann. infer. **90**(2), 227–244 (2000)
7. Liao, X., Xue, Y., Carin, L.: Logistic regression with an auxiliary data source. In: Proceedings of the 22nd International Conference on Machine learning. ACM (2005)
8. Rosenstein, M.T., et al.: To transfer or not to transfer. In: NIPS 2005 Workshop on Transfer Learning, vol. 898 (2005)
9. Dai, W., et al.: Boosting for transfer learning. In: Proceedings of the 24th International Conference on Machine Learning. ACM (2007)
10. Freund, Y., Schapire, R.E.: A desicion-theoretic generalization of on-line learning and an application to boosting. In: Vitányi, P. (ed.) EuroCOLT 1995. LNCS, vol. 904, pp. 23–37. Springer, Heidelberg (1995)
11. Blum, A., Mitchell, T.: Combining labeled and unlabeled data with co-training. In: Proceedings of the Eleventh Annual Conference on Computational Learning Theory. ACM (1998)
12. Zadrozny, B.: Learning and evaluating classifiers under sample selection bias. In: Proceedings of the Twenty-First International Conference on Machine Learning. ACM (2004)
13. Huang, J., et al.: Correcting sample selection bias by unlabeled data. In: Advances in Neural Information Processing Systems (2006)
14. Kohavi, R.: A study of cross-validation and bootstrap for accuracy estimation and model selection. In: Ijcai, vol. 14, no. 2 (1995)
15. Sugiyama, M., et al.: Direct importance estimation with model selection and its application to covariate shift adaptation. In: Advances in Neural Information Processing Systems (2008)
16. Li, B., Yang, Q., Xue, X.: Transfer learning for collaborative filtering via a rating-matrix generative model. In: Proceedings of the 26th Annual International Conference on Machine Learning. ACM (2009)
17. Cleuziou, G.: An extended version of the k-means method for overlapping clustering. In: 19th International Conference on Pattern Recognition, ICPR 2008. IEEE (2008)
18. Park, Y.-J., Tuzhilin, A.: The long tail of recommender systems and how to leverage it. In: Proceedings of the 2008 ACM conference on Recommender systems. ACM (2008)
19. Hotho, A., Steffen, S., Stumme, G.: Ontologies improve text document clustering. In: Third IEEE International Conference on Data Mining, ICDM 2003. IEEE (2003)
20. Dhillon, I.S., Mallela, S., Modha, D.S.: Information-theoretic co-clustering. In: Proceedings of the Ninth ACM SIGKDD International Conference on Knowledge Discovery and Data Mining. ACM (2003)

Robust Modified ABC Variant (JA-ABC5b) for Solving Economic Environmental Dispatch (EED)

Noorazliza Sulaiman[1], Junita Mohamad-Saleh[1(✉)], and Abdul Ghani Abro[2]

[1] School of Electrical and Electronic Engineering,
Universiti Sains Malaysia, Nibong Tebal, Malaysia
noorazlizasulaiman@gmail.com, jms@usm.my
[2] College of Applied Engineering, King Saud University,
Al Muzahimiyah Campus, Riyadh, Kingdom of Saudi Arabia
ghaniabro@gmail.com

Abstract. Artificial bee colony (ABC) algorithm has been widely used to solve various optimization problems due to its simplicity and flexibility besides showing outrageous results in comparison to other optimization algorithms. Nevertheless, ABC has been found to suffer from few limitations such as slow convergence rates and premature convergence tendency. With the motivation to overcome the problem, this work proposes a modified ABC variant referred to as JA-ABC5b with the aim of robust and faster convergence. The proposed ABC variant has been compared with the standard ABC and other existing ABC variants on 27 benchmarks functions and to solve economic environmental dispatch (EED) problem. The results have shown that JA-ABC5b has the best performance in comparison to the standard ABC and selected existing ABC variants in terms of convergence speed as well as the global optimum achievement besides exhibiting robust performance in solving complex real-world optimization problem.

Keywords: Artificial bee colony · ABC variants · Swarm-intelligence-based algorithm · Convergence speed · Benchmark functions · Economic environmental dispatch

1 Introduction

Bio-inspired algorithms (BIAs), inspired by the behaviors of nature have been applied to solve various complex optimization problems [1] as shown by the works of [2–5]. They have been implemented to solve the concern of various problems i.e. the problems of high computational cost and premature convergence tendency faced by numerical methods [4]. The outcomes are the promising results of BIAs at solving those problems [6].

BIAs are metaheuristic method consisting of several classes and one of the most prominent classes is swarm-intelligence-based (SI) algorithms. SI algorithms have been found to show tremendous performance in solving various problems such as the

© Springer International Publishing Switzerland 2016
H. Ohwada and K. Yoshida (Eds.): PKAW 2016, LNAI 9806, pp. 55–67, 2016.
DOI: 10.1007/978-3-319-42706-5_5

travelling salesman problem [7], power loss minimization [8], voltage profile improvement [9], economic environmental dispatch [10, 11], stability enhancement for multi-machine power system [12] and many more. A few examples of SI algorithms are ant colony optimization (ACO) [13], particle swarm optimization (PSO) [14] and artificial bee colony (ABC) algorithms [15]. Artificial bee colony (ABC) algorithm, one of the SI methods has recently attracted the attention of optimization researchers due to its records of outrageous performances [16–18].

Inspired from the foraging behavior of honeybees, ABC was proposed by Karaboga in 2005 [15]. It is basically a type of computational method [16]. Besides showing efficiency, ABC has demonstrated excellent performance in comparison to other prominent optimization algorithms such as genetic algorithm (GA), differential evolution (DE), particle swarm optimization (PSO) algorithms and few others [16–18] at solving various optimization problems. Apart from that, its controlled parameters are easy to be tuned and the implementation of its algorithm is simple and flexible [16]. Hence, many optimization researchers have shown their interest in this algorithm.

Nevertheless, ABC suffers from few limitations, i.e. slow convergence speed on unimodal functions and premature convergence tendency on multimodal functions caused by the local minima trappings [19, 20]. Basically, those limitations are due to the mutation equation of ABC which is good in exploration but poor in exploitation. Moreover, it portrays extreme self-reinforcement that makes ABC exhibits less capabilities in exploitation. Hence, these have motivated many researchers to propose various ABC variants [19–25] with the aims to solve these problems. Nonetheless, none of them are able to solve both problems simultaneously [19, 20, 22]. The imbalanced exploration and exploitation processes in the standard ABC have been found to be the main reason for its lack of performance and robustness [23–25]. Thus, this work proposes a modified ABC variant named JA-ABC5b which is an extended version of the proposed ABC variant found in [26]. It serves to balance out the exploration and exploitation capabilities and is expected to persistently avoid local optima traps and converge faster.

2 Modified ABC Variant (JA-ABC5b)

In the standard ABC algorithm, its mutation equation works by directing the interaction between the food source to be updated and a randomly chosen food source, which means that it is chosen regardless of its fitness value. Hence, in any case, if fitter random food source is selected, the candidate solution would be fitter as well. The problem arises when the food source with poor fitness values has been randomly chosen for the interaction. The produced candidate solution would be directed close to the poor possible solution. In other words, the candidate solution would be poor and drifted away from the global minimum.

To overcome the problem, one previous work [26] suggested few modifications to the standard ABC algorithm. The extension of this previous work has produced JA-ABC5b. The working procedures of JA-ABC5b are illustrated in Fig. 1. The modifications done to the standard ABC are highlighted in the figure. The first

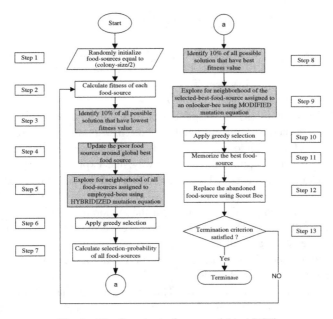

Fig. 1. The flowchart of proposed JA-ABC5b

modification is the insertion of new stages at the earlier stage of the standard ABC. The first step in the new stages is to identify the poor food sources.

Next, the poor food sources are being updated around global best (g-best) solution using the mutation equation inspired from [23]:

$$z_{ij} = y_{best,j} + \varphi_{ij}(y_{pj} - y_{kj}) \tag{1}$$

where z_{ij} represents the candidate solution of i-th food source with j-th dimension. $y_{best,j}$ is the best food source, y_{pj} represents j-th dimension of p-th food source and is randomly chosen. Subscripts i, k and p are mutually exclusive food sources and ϕ_{ij} is a random number within $[-1, 1]$, respectively.

Equation (1) directs the food sources towards the best solutions. Hence, the resultant candidate solution is expected to be fitter than the previous ones and will substitute the poor identified food sources.

Essentially, this action increases the number of fitter food sources in the population and thus, create fitter overall population. Next, employed-bees update those food sources. Thus, the convergence speed of the algorithm increases because the food sources have already been guided towards the best solutions.

However, since the population is now fitter, the search space might become more complex and thus, there is a possibility for the algorithm to be trapped in local optima. To overcome this, random exploration is required for the employed-bees to avoid local optima trappings. The mutation equation of the employed-bees has been modified for the purpose. The modification is done by hybridizing the mutation equations found in [22, 27] to become:

$$z_{ij} = y_{r1,j} + \varphi_{ij}(y_{r2,j} - y_{r3,j}) + \psi_{ij}(y_{r4,j} - y_{best,j}) \tag{2}$$

where $y_{r1,j}$, $y_{r2,j}$, $y_{r3,j}$ and $y_{r4,j}$ are $r1$, $r2$, $r3$ and $r4$-th food sources with j-th dimension. Subscripts $r1$, $r2$, $r3$ and $r4$ refer to the mutually exclusive food sources, Ψ_{ij} is a random number within $[0,T]$, where T is a user defined number and the rest of the parameter are the same as (1).

The first two terms of (2) are inspired from the equation proposed in [22]. The equation has been modified prior to hybridizing it with the other equation. This equation is known for its randomness [22] and thus it has been further enhanced with the aim to improve its random search. This way, the exploration capability of the algorithm has also been enhanced. This action will avoid the possibility for premature convergence. The last term of the equation is inspired from the work of [27]. The insertion of the term is targeted to balance out the exploration and exploitation capabilities of the algorithm [27]. The overall equation is expected to simultaneously exhibit faster convergence speed as well as efficient local minima avoidance.

The next modification is to make sure that the overall algorithm consistently balances out exploration and exploitation capabilities. This is done by enhancing the exploitation capability of the algorithm. The onlooker-bees have been directed to update only few (10 %) most-fit selected-food sources instead of updating all selected-food sources. With only few food sources to be updated, the exploitation coverage of onlooker-bees becomes smaller and hence, the algorithm can converge faster.

The next modification is to replace the mutation equation of the onlooker-bees with the mutation equation found in [22]. The replacement equation is given by:

$$z_{ij} = y_{best,j} + \varphi(y_{ij} - y_{kj}) \tag{3}$$

where z_{ij} is the candidate solution of food sources and $y_{best,j}$ is the best food source in the current population. The rest of the parameters are the same as (1).

Equation (3) basically has the ability to increase the exploitation process of onlooker-bees as it exploits the information of best food source to update the selected fitter-food sources. With less number of food sources to be updated in onlooker-bees phase, the algorithm should be able to converge faster. Also, fitter food sources lead towards the global optimum solution. Thus, JA-ABC5b is able to balance the exploration and exploitation capabilities and be able to converge faster in finding global optimum.

3   Economic Environmental Dispatch (EED)

Economic environmental dispatch (EED) is a power systems-based problem that relates to the emission of toxic gases from fossil fuels as a result of combustion of fossil fuels during electricity generation. The effects of combustion fossil fuels in generating electricity create the worst pollution in major cities. The emission of toxic gases into the atmosphere has become a concern to many people due to its hazardous effects. Hence, an efficient solution is vital to solve the problems [28–30].

Basically, EED serves to determine the committed generating units in order to meet the demand while utilizing minimum operating cost and emitting minimum toxic gases. All the matters above are subjected to the system constraints such as power balance and active power lower and upper limit [29, 30]. EED is known to be a constrained-based multi-objective problem [10]. Moreover, the objectives of the problem are contradicted to each other, making the problem to be complex and hard to solve.

In the effort to solve it, researchers have implemented various types of techniques such as goal-programming techniques [31], multi-objective differential evolution [10] and many more in order to solve this problem. However, most of the techniques have treated the problem as a single objectives problem, requiring assumptions to be made [32]. In the meantime, there are techniques that evaluated EED as multi-objective problem, but they have been found to be computationally intensive and time consuming [10, 33]. This is due to the complexity of EED as a multi-objective optimization problem.

The mathematical formulation of all the problem's objectives and its constraints are being formulated in the following subsection.

3.1 Cost

One of EED's objectives is to minimize the operating cost. The mathematical formulation of the operating cost after taking valve point effect into consideration is as follows:

$$F_i(P_i) = \sum [a_i P_i + b_i P_i + c_i + |d_i \sin\{e_i(P_i^{\min} - P_i)\}|] \qquad (4)$$

where P_i is the power generated by i-th generator unit, a_i, b_i and c_i are the cost coefficients of i-th generator unit, d_i and e_i are the fuel coefficients of i-th generator unit that show the valve point effect [34, 35] and N is the total number of generator unit. Valve point effect basically occurs due to the effect of wire drawing. It happens when each of the steam admission valves starts to open, resulting in a ripple that can cause the objective function to have higher order nonlinearity and become non-smooth [35].

3.2 Emission

The second objective of the EED problem is to minimize the emission of the hazardous gases into the atmosphere. The mathematical formulation of the emission is given by:

$$F_i(P_i) = \sum_{i=1}^{N} [\gamma_i P_i^2 + \beta_i P_i + \alpha_i + |\eta_i \exp(\delta_i P_i)|] \qquad (5)$$

where P_i is the power generated by i-th generator unit, α_i, β_i, γ_i, δ_i and η_i are the emission coefficients of i-th generator unit and N is the total number of generator unit.

Equation (5) expresses the total summation of the toxic gases released by the generators. The toxic gases include sulfur oxides (SO_x), nitrogen oxides (NO_x) and carbon dioxide (CO_2) which can cause harm to living this. Thus, the emission of those gases needs to be minimized.

3.3   Constraints

To minimize both objective functions i.e. operating cost and gases emission, the control variables of the system need to be optimized within their limits and to satisfy the equality and inequality constraints. In EED problem, the control variables which are the power generated by each of the committed generating units need to satisfy the constraints that are being presented in the following subsection.

3.3.1   Equality Constraint

The equality constraint of EED is active power balance. Thus, the equation of the active power balance is given by:

$$\sum_{i=1}^{N} P_i - P_D - P_L = 0 \tag{6}$$

where P_i is the power generated by i-th generator unit, P_D is the active power demand and P_L is the transmission line losses. The power generated by each of the committed generator unit has to satisfy this active power balance.

3.3.2   Inequality Constraints

The active power generated at each are committed generating units needs to be limited within the inequality constraints, which are the upper and lower limit. The limit is as follows:

$$P_i^{min} \leq P_i \leq P_i^{max} \quad t = 1, \ldots, N \tag{7}$$

where P_i^{min} is the lower limit of active power generated by i-th generator unit, P_i^{max} is the upper limit of active power generated by the i-th generator unit and the rest of the parameters are similar to (4).

3.4   Penalty Function

Penalty function is derived prior to converting a multi-objective function with equality and inequality constraints into a single objective function. Penalty terms have been added to the objective function for that purpose. Thus, the objective function in the form of penalty function of EED is given by:

$$P(x) = \sum_{i=1}^{N} F_i(P_i) + \sum_{i=1}^{N} E_i(P_i) + K \times \left\{ \left(\sum_{i=1}^{N} P_i \right) - P_L - P_D \right\}^2 \tag{8}$$

where F_i is the cost function, E_i is the emission function, P_i is the active power generated at i-th generator unit, K is the penalty term of the penalty function, P_L is the power losses of the transmission lines and P_D is the power demand. Equation (8) will be used as the objective function to be solved by the proposed JA-ABC5b algorithm.

More details of the problem formulation can be found in [10, 35].

4 Experimental Setup

For performance evaluation, the proposed algorithm, JA-ABC5b has been simulated on twenty-seven commonly used benchmark functions as listed in Table 1. Its performance has been compared with the standard ABC (ABC) [16] and the three best-performed existing ABC variants, i.e. global best ABC (BABC1) [23], improved ABC (IABC) [22] and enhanced ABC (EABC) [27]. For all compared algorithms, the dimension of the benchmark function has been set to 30, the population size has been set to 50 and number of generation has been limited to 1000. As for optimizing the EED problem, 10-unit generator system has been used. The dataset can be found in [10]. The performance of JA-ABC5b in solving EED problem has been compared with the standard ABC, BABC1, IABC, EABC as well as with other optimization algorithms available in the work of [10].

Table 1. Benchmark functions

Function	Function name	Initialization range
f1	Griewank	±600
f2	Rastrigin	±15
f3	Rosenbrock	±15
f4	RS Ackley	±32
f5	Schwefel	±500
f6	Himmelblau	±600
f7	RS Sphere	±600
f8	Step	±600
f9	Bohachevsky 2	±100
f10	RS Schwefel 2.22	±100
f11	RS Schwefel Ridges	±100
f12	RS Schwefel Ridges with Noise	±15
f13	RS Elliptic	±100
f14	Zekhelip	±15
f15	Non-continuos Rastrigin	±15
f16	Michalewicz	0–180
f17	First Expanded Function	±15
f18	Second Expanded Function	±15
f19	Third Expanded Function	±15
f20	Fourth Expanded Function	±500

(Continued)

Table 1. (*Continued*)

Function	Function name	Initialization range
f21	Fifth Expanded Function	±*100*
f22	Sixth Expanded Function	±*100*
f23	Seventh Expanded Function	±*15*
f24	Eighth Expanded Function	±*100*
f25	Rotated Griewank Function	*0–600*
f26	Rotated Ackley Function	±*32*
f27	Rotated Rastrigin Function	±*5*

For global solution validation, each of the compared algorithms including JA-ABC5b has been set to run for 30 times on each benchmark function [19]. This repetition is necessary to ensure that the algorithm does not get trapped in a local minimum. All the values used in the experiments follow those used and recommended in literature works [16, 17, 19, 22–25]. The simulation and testing process have been done using Matlab 2010a programming language on Intel Core i7 CPU 2.80 GHz computer.

5   Results and Discussions

Figures 2, 3, 4, 5 and 6 show the performance results of the proposed JA-ABC5b in comparison with the standard ABC and the existing ABC variants i.e. BABC1, IABC and EABC.

The results have shown the superiority of JA-ABC5b in comparison to the compared algorithms on most of the benchmark functions, particularly on *f1*, *f4*–*f16*, *f18*–*f19*, *f21*–*f25* and *f27*. The results have been evaluated on the basis of convergence speed and global minimum achievement.

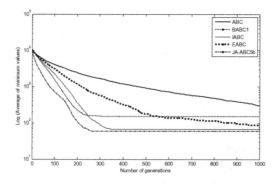

Fig. 2. Performance result of JA-ABC5b on *f5*

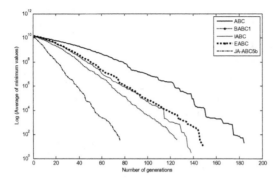

Fig. 3. Performance result of JA-ABC5b on *f*6

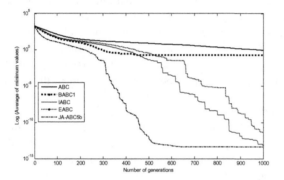

Fig. 4. Performance result of JA-ABC5b on *f*15

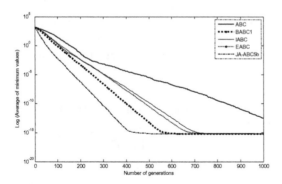

Fig. 5. Performance result of JA-ABC5b on *f*18

Figures 2, 4 and 6 have depicted the inefficiency of the compared algorithm particularly ABC, BABC1 and EABC in achieving global optimum on *f*5, *f*15 and *f*27. Meanwhile, JA-ABC5b has diligently obtained minimum solution on those functions.

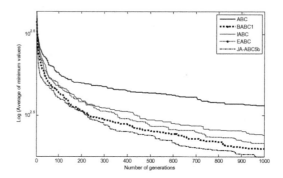

Fig. 6. Performance result of JA-ABC5b on *f*27

Although IABC and EABC did reach the global optimum on *f*15 but JA-ABC5b has shown faster convergence speed as compared to them.

Moreover, JA-ABC5b has efficiently reached the global minimum for *f*5 whilst other algorithms get trapped in local optima. In the meantime, JA-ABC5b has exhibited faster convergence speed than the other algorithms on the rest of the graphs. The enhancement of the exploitation process of the algorithm in the onlooker-bees phase has speed-up the convergence rates of the algorithm. Thus, JA-ABC5b is able to perform better than the others, as illustrated by the figures.

Besides that, the performance of JA-ABC5b has also been compared with other optimization algorithms: multi-objective differential evolution (MODE), pareto differential evolution (PDE), nondominated sorting genetic algorithm-II (NSGA-II) and strength pareto evolutionary algorithm 2 (SPEA2) algorithms. The results are presented in Table 2. The results clearly show that JA-ABC5b requires the least amount of operating cost and emits least toxic gases in comparison to other ABC variant and optimization algorithms when generating electricity.

Based on the performance results obtained, it can be concluded that the proposed ABC algorithm, JA-ABC5b exhibits the best performance among all variants. Hence, it can be regarded as a reliable optimization algorithm for solving a complex optimization problem.

Table 2. Performance results of compared optimization algorithms in minimizing EED on 10-Unit generator system

Algorithm	EED (10^5)($/(lb))
MODE	1.1760
PDE	1.1762
NSGA-II	1.1767
SPEA2	1.1763
ABC	1.1737
BABC1	1.1686
IABC	1.1693
EABC	1.1692
JA-ABC5b	**1.1677**

6   Conclusion

The work proposes a modified ABC variant referred to as JA-ABC5b which has been developed with the objectives of improving the convergence speed and diligently avoiding local minima traps. Few modifications have been done to the standard ABC algorithm to balance the exploration and exploitation capabilities of the standard ABC algorithm, producing a new ABC variant, JA-ABC5b. The performances of JA-ABC5b have been compared with the existing ABC variants and the results obtained have clearly proved the excellent performance of the JA-ABC5b as on various commonly used benchmark functions. This verifies better convergence speed and local minima avoidance of the algorithm.

For further analysis, JA-ABC5b has been implemented to optimize the economic environmental dispatch (EED) problem. The performance of JA-ABC5b in solving the problem has been observed and compared with the existing ABC variants as well as the other optimization algorithms. JA-ABC5b has shown outrageous results in solving this complex real-world problem in comparison to other algorithms. The results have vividly suggested that the proposed algorithm is a robust optimization algorithm which shows the ability to solve complex optimization problem.

Acknowledgments. The authors acknowledge the Ministry of Higher Education (MOHE) FRGS Grant No: 203/PELECT/6071247 and the Universiti Sains Malaysia (USM) for the financial and facilities support.

References

1. Layeb, A.: A hybrid quantum inspired harmony search algorithm for 0–1 optimization problems. J. Comput. Appl. Math. **253**, 14–25 (2013)
2. Hendrawan, Y., Murase, H.: Bio-inspired feature selection to select informative image features for determining water content of cultured Sunagoke moss. Expert Syst. Appl. **38**(11), 14321–14335 (2011)
3. Kurid, H.A., Alnusairi, T.S., Almujahed, H.S.: OBAME: optimized bio-inspired algorithm to maximize search efficiency in P2P databases. Procedia Comput. Sci. **21**, 60–67 (2013)
4. Rocha, M., et al.: Optimization of fed-batch fermentation processes with bio-inspired algorithms. Expert Syst. Appl. **41**(5), 2186–2195 (2014)
5. Youssef, B.B.: Parallelization of a bio-inspired computational model for the simulation of 3-D multicellular tissue growth. Procedia Comput. Sci. **20**, 391–398 (2013)
6. Binitha, S., Sathya, S.S.: A survey of bio inspired optimization algorithms. Int. J. Soft Comput. Eng. **2**(2), 2231–2307 (2012)
7. Ding, C., Cheng, Y., He, M.: Two-level genetic algorithm for clustered traveling salesman problem with application in large-scale TSPs. Tsinghua Sci. Technol. **12**(4), 459–465 (2007)
8. Tehzeeb-Ul-Hassan, H., et al.: Reduction in power transmission loss using fully informed particle swarm optimization. Int. J. Electr. Power Energy Syst. **43**(1), 364–368 (2012)
9. Musirin, I., et al.: Voltage profile improvement using unified power flow controller via artificial immune system. WSEAS Trans. Power Syst. **3**(4), 194–204 (2008)
10. Basu, M.: Economic environmental dispatch using multi-objective differential evolution. Appl. Soft Comput. **11**(2), 2845–2853 (2011)

11. Kumar, K.S., et al.: Economic load dispatch with emission constraints using various PSO algorithms. WSEAS Trans. Power Syst. **3**(9), 598–607 (2008)

12. Oonsivilai, A., Marungsri, B.: Stability enhancement for multi-machine power system by optimal PID tuning of power system stabilizer using particle swarm optimization. WSEAS Trans. Power Syst. **3**(6), 465–474 (2008)

13. Dorigo, M., Maniezzo, V., Colorni, A.: Ant system: optimization by a colony of cooperating agents. IEEE Trans. Syst. Man Cybern. Part B Cybern. **26**(1), 29–41 (1996)

14. Kennedy, J., Eberhart, R.: Particle swarm optimization. In: Proceedings of IEEE International Conference on Neural Networks, Perth, pp. 1942–1948 (1995)

15. Karaboga, D.: An Idea Based on Honey Bee Swarm for Numerical Optimization. Technical report (2005)

16. Karaboga, D., Akay, B.: A comparative study of artificial bee colony algorithm. Appl. Math. Comput. **214**(1), 108–132 (2009)

17. Karaboga, D., Basturk, B.: A powerful and efficient algorithm for numerical function optimization: artificial bee colony (ABC) algorithm. J. Global Optim. **39**(3), 459–471 (2007)

18. Karaboga, D., Basturk, B.: On the performance of artificial bee colony (ABC) algorithm. Appl. Soft Comput. **8**(1), 687–697 (2008)

19. Abro, A.G., Mohamad-Saleh, J.: Enhanced global-best artificial bee colony optimization algorithm. In: Sixth UKSim/AMSS European Symposium on Computer Modeling and Simulation, Valetta, pp. 95–100 (2012)

20. Abro, A.G., Mohamad-Saleh, J.: An enhanced artificial bee colony optimization algorithm. In: Mastorakis, N., Simian, D., Prepelita, V. (eds.) Recent Advances in Systems Science and Mathematical Modelling, pp. 222–227. WSEAS Press (2012)

21. Banharnsakun, A., Achalakul, T., Sirinaovakul, B.: The best-so-far selection in artificial bee colony algorithm. Appl. Soft Comput. **11**(2), 2888–2901 (2011)

22. Gao, W., Liu, S.: Improved artificial bee colony algorithm for global optimization. Inf. Process. Lett. **111**(17), 871–882 (2011)

23. Gao, W., Liu, S., Huang, L.: A global best artificial bee colony algorithm for global optimization. J. Comput. Appl. Math. **236**(11), 2741–2753 (2012)

24. Gao, W.-F., Liu, S.-Y.: A modified artificial bee colony algorithm. Comput. Oper. Res. **39**(3), 687–697 (2012)

25. Zhu, G., Kwong, S.: Gbest-guided artificial bee colony algorithm for numerical function optimization. Appl. Math. Comput. **217**(7), 3166–3173 (2010)

26. Sulaiman, N., Mohamad-Saleh, J.: Modified ABC variant (JA-ABC4) for performance enhancement. In: International Conference on Mathematical Methods, Mathematical Models and Simulation in Science and Engineering, Interlaken, pp. 178–183 (2014)

27. Abro, A.G.: Performance Enhancement of Artificial Bee Colony Optimization Algorithm. Ph.D. thesis, Universiti Sains Malaysia (2013)

28. Breeze, P.: Power Generation Technologies. Elsevier Science, United Kingdom (2005)

29. Po-Hung, C., Hong-Chan, C.: Large-scale economic dispatch by genetic algorithm. IEEE Trans. Power Syst. **10**(4), 1919–1926 (1995)

30. Sinha, N., Chakrabarti, R., Chattopadhyay, P.K.: Evolutionary programming techniques for economic load dispatch. IEEE Trans. Evol. Comput. **7**(1), 83–94 (2003)

31. Nanda, J., Kothari, D.P., Lingamurthy, K.S.: Economic-emission load dispatch through goal programming techniques. IEEE Trans. Energy Convers. **3**(1), 26–32 (1988)

32. Farag, A., Al-Baiyat, S., Cheng, T.C.: Economic load dispatch multiobjective optimization procedures using linear programming techniques. IEEE Trans. Power Syst. **10**(2), 731–738 (1995)

33. Das, D.B., Patvardhan, C.: New multi-objective stochastic search technique for economic load dispatch. IEEE Proc. Gener. Transm. Distrib. **145**(6), 747–752 (1998)

34. Borckmans, P.B., et al.: A Rienman subgradient algorithm for economic dispatch with valve-point effect. J. Comput. Appl. Math. **255**, 848–866 (2014)
35. Coelho, L.D.S., Mariani, V.C.: An efficient cultural self-organizing migrating strategy for economic dispatch optimization with valve-point effect. Energy Convers. Manage. **51**(12), 2580–2587 (2010)

Knowledge Acquisition and Natural Language Processing

Enhanced Rules Application Order to Stem Affixation, Reduplication and Compounding Words in Malay Texts

Mohamad Nizam Kassim[1,2(✉)], Mohd Aizaini Maarof[2],
Anazida Zainal[2], and Amirudin Abdul Wahab[1]

[1] Strategic Research, CyberSecurity Malaysia,
The Mines Resort City, 43300 Seri Kembangan, Malaysia
{nizam, amirudin}@cybersecurity.my
[2] Faculty of Computing, Universiti Teknologi Malaysia,
81310 Skudai, Johore, Malaysia
{aizaini, anazida}@utm.my

Abstract. Word stemmer is an automated program to remove affixes, clitics and particles from derived words based on morphological structures of specific natural languages. It has been widely used for text preprocessing in many artificial intelligence applications. Furthermore, the performance of word stemmer to correctly stem derived words has an influence to the performance of information retrieval, text mining and text categorization applications. Despite of various stemming approaches were proposed in the past research, the existing word stemmers for Malay language still suffer from stemming errors. Moreover, the existing word stemmers partially consider morphological structures of Malay language in which only focused on affixation words instead of affixation, reduplication and compounding words, simultaneously. Therefore, this paper proposes an enhanced word stemmer using rule-based affixes removal and dictionary lookup methods called enhanced rule application order that is able to stem affixation, reduplication and compounding words and at the same time, is able to address possible stemming errors. This paper also examines possible root causes of affixation, reduplication and compounding stemming errors that could happen during word stemming process. The experimental results indicate that the proposed word stemmer is able to stem affixation, reduplication and compounding words with better stemming accuracy by using enhanced rule application order.

Keywords: Malay word stemmer · Word stemming algorithm · Word stemmer · Stemming error · Rules application order

1 Introduction

Generally, the term word stemming in computational linguistics refers to the process of reducing derived words to their respective root words by removing inflectional and derivational affixes, clitics and particles based on morphological structures of specific natural language [2]. Word stemmer is an automated program that has been widely

© Springer International Publishing Switzerland 2016
H. Ohwada and K. Yoshida (Eds.): PKAW 2016, LNAI 9806, pp. 71–85, 2016.
DOI: 10.1007/978-3-319-42706-5_6

used to perform word stemming process before applying statistical computation methods in artificial intelligence domains such as information retrieval, text mining, knowledge management and text categorization [23].

Julie Beth Lovins developed the first word stemmer for English language that had great impact to the later research on the subsequent word stemmers in other natural languages [15, 17]. The development of word stemmer heavily depends on the morphological structures of specific natural languages such as language characters and word patterns [3]. For instance, word stemmer for English language is considerably sufficient to remove only suffixes e.g. the derived words of '*confuses*', '*confusing*', '*confused*' and '*confusion*', which are derived from the root word '*confuse*'. In contrast, word stemmer for Malay language requires more complex affixes removal method in order to remove prefixes, suffixes, confixes and infixes e.g. the derived words of '*menyambung*', '*sambungan*', '*penyambungan*' and '*sinambung*' which are derived from the root word '*sambung*'. Therefore, suffixes removal alone is not sufficient to identify the correct root word in Malay language [2]. It requires the understanding on how derived words are formed from their respective root words.

There are many word stemmers for Malay language have been developed for various applications such as general purpose [12, 19], information retrieval [1, 2, 7, 8, 14, 16, 20, 24], automated essay grading system [11] and text categorization [25]. However, these word stemmers have two word stemming limitations in addressing the morphological structures of Malay language.

The first word stemming limitation relates to partial consideration on the morphological structures of Malay language. Generally, there are three different derived words in Malay language namely affixation, reduplication and compounding words but the existing word stemmers only focused on affixation words in their word stemming rules. To date, none of the existing word stemmers is able to stem affixation, reduplication and compounding words, simultaneously.

The second word stemming limitation relates to the challenges in addressing existing stemming errors to identify the correct root words namely overstemming, understemming, unstem, and special variation and exception errors [1, 12–14, 25]. It is due to affixes, clitics, and/or particles removal from derived words require a solid understanding on how derived words are formed from combination of affixes, clitics, and/or particles with root words.

The understanding of these morphological structures provides a good foundation for developing word stemming rules based on the morphological structures of Malay language. Therefore, this paper proposes an enhanced word stemmer for Malay language using enhanced rules application order that is able to stem affixation, reduplication and compounding words while addressing possible affixation, reduplication and compounding stemming errors during word stemming process.

Hence, this paper is organized into six subsequent sections. Section 2 discusses word morphology in Malay language. Section 3 describes the related works in Malay word stemming. Section 4 examines word patterns and their respective stemming errors. Section 5 describes the proposed word stemmer that stems affixation, reduplication and compounding words. Section 6 discusses the experimental results where Malay online news articles have been used to evaluate the proposed stemmer. Finally, Sect. 7 concludes this paper.

2   Word Morphology in Malay Language

Malay language (*Bahasa Melayu*) is an Austronesian family which is the most spoken language in the Malay Archipelago (Malaysia, Indonesia, Singapore and Brunei) [10, 18]. Malay language is also known with various official names in which depending on geographical location such as *Bahasa Malaysia* in Malaysia, *Bahasa Indonesia* in Indonesia and *Bahasa Melayu* in Singapore and Brunei [10]. In Malaysia, *Dewan Bahasa dan Pustaka* (Council of Language and Literary) is a government agency that holds responsibility for creating new words and managing root words in Malay language [20].

The Malay morphology is known to have very complex morphological structures to form various word patterns. There are seven word patterns in Malay language namely affixation, reduplication, compounding, blending, clipping, abbreviation and borrowing [10].

Affixation words are derived words from the combination of affixes, clitics and particles with root words e.g. *memakan* (to eat), *makanan* (foods), *pemakanan* (nutrition). Affixes are morphemes that identify part-of-speech of the words to represent nouns, verbs or adjectives. Unlike English language, affixes are used to represent plural, tenses and possession of the words. Furthermore, there are two different types of affixes namely inflectional affixes and derivational affixes. Inflectional affixes do not change part-of-speech of the root words e.g. *hendak* (want) → *dikehendaki* (wanted) whereas derivational affixes do change part-of-speech of the root words e.g. *makan* (eat) → *pemakanan* (nutrition). On other hand, reduplication words are one of derived words in Malay language in which reflect the plural form by repeating the words with hyphen (-) e.g. *burung-burung* (birds). In some instances, reduplication words contain affixes, clitics or particles e.g. *pelajar-pelajar* (students) and *surat-menyurat* (correspondence). Compounding words are another word patterns that combined two or three root words to form derived words e.g. *ambil* (to take) and *alih* (to remove) → *ambilalih* (to takeover).

Only affixation, reduplication and compounding words are considered as derived words in Malay language whereas other word patterns such as blending, clipping, abbreviation and borrowing are considered as root words. These words can be described as following examples i.e. abbreviation e.g. *orang* (people) → *org* (people), blending e.g. *cerita pendek* (short story) → *cerpen* (short story), clipping e.g. *emak* (mother) → *mak* (mother) and borrowing e.g. *kontemporari* (contemporary). It is important to note that every natural language such as Malay language, English language and Arabic language has their own morphological structures to describe their own word patterns. For instance, Malay language does not have grammatical function referring to gender, tenses or verbs whereas English language and Arabic language have these morphological rules in their respective languages.

Therefore, the understanding of morphological structures of Malay language is very important for developing word stemmer for Malay language due to the goal of word stemmer is to reduce derived words into their respective root words. These derived words are affixation, reduplication and compounding words that need to be stemmed into their respective root words.

2.1   Affixation Words and Their Respective Morphological Structures

Affixation words may comprise of eleven possible combinations of root words with affixes, clitics and/or particles to form subcategories of affixation words as described in Table 1. Affixation words can be categorized into four different word patterns namely prefixation, suffixation, confixation, and infixation words. Prefixation words may contain single or multiple prefixes (usually attached at the beginning of root words) with root words e.g. *perompak* (robber), and *memperdaya* (to scam). Suffixation words may contain single suffixes or combination of suffixes and enclitics/particles with root words (usually attached at the ending of root words) e.g. *langganan* (subscription) and *berikanlah* (just give). Confixation words may contain confixes (also known as cir-cumfixes and prefix-suffix pairs) with root words in which single or multiple prefixes and sometime, proclitics (usually attached at the beginning of the root word) and single suffixes or combination of suffixes/enclitics/particles (usually attached at the ending of the root word) simultaneously e.g. *bergembiralah* (have fun), *pengetahuan* (knowl-edge), *kelakiannya* (his manhood) and *dipermudahkan* (to be ease). Infixation words may contain infixes that are attached at the middle of the root words *telunjuk* (pointing fingers), *sinambung* (continuity) and *gemuruh* (feel fear). There are also special vari-ation and exception rules in deriving prefixation and confixation words whereby pre-fixes e.g. *meny+*, *pem+* and *pen+* require first letter of root word to be dropped e.g. *meny + sapu* $\rightarrow$ *menyapu*, *pem + fikir* (think) $\rightarrow$ *pemikir* (thinker) and *pen + tolong* (assist) $\rightarrow$ *penolong* (assistant).

Table 1. Eleven possible combinations of affixation words.

Affixation	Combination of root words with affixes, clitics and particles
Prefixation	1. prefix+root word: [*perompak* (robber)] 2. multiple prefixes + root word: [*memperdaya* (to scam)]
Suffixation	1. root word + suffix: [*langganan* (subscription)] 2. root word+suffix+enclitic/particles [*berikanlah* (just give)]
Confixation	1. prefix + root word + suffix: [*pengetahuan (knowledge)*] 2. prefix + root word + enclitics/particle: [*bergembiralah* (have fun)] 3. multiple prefixes+root word+suffix: [*memperjudikan* (gamble)] 4. prefix + root word + suffix + enclitic/particles [*kelakiannya* (his manhood)] 5. multiple prefixes + root word + suffix + enclitic/particles: [*mempertemukannya* (his/her findings)] 6. proclitic + root word + suffix: *kunantikan* (I'm waiting)]
Infixation	1. root word with infix: [*telunjuk* (pointing fingers)]

2.2   Reduplication Words and Their Respective Morphological Structures

Reduplication words are derived words that function as plural form or adjective of the root words. There are two types of word patterns in reduplication words. The first word pattern of reduplication word contains two words with hyphen (-). These reduplication words may be repeating words with hyphen (-) e.g. *gunung-gunung* (mountains), repeating words with hyphen (-) but one of them is in the form of rhythmic word e.g.

warna-warni (colourful), and repeating words with hyphen (-) but one of them is affixation word e.g. *surat-menyurat* (correspondence). The second word pattern of reduplication words is reduplication words without hyphen (-) but additional syllable is attached to the root words. For instance, the word *tetamu* (visitor) is derived from the root word, *tamu* (visit) and the syllable *te+*, which is not affixes, clitics or particles. Thus, there are thirteen different word patterns of reduplication words namely full reduplication, rhythmic reduplication, affixed reduplication and partial reduplication as described in Table 2.

Table 2. Thirteen possible combinations of reduplication words.

Reduplication	Combination of root words with affixes and hyphen
Full Reduplication	*gunung* (mountain) → *gunung-gunung* (mountains)
Rhythmic Reduplication	*warna* (colour) → *warna-warni* (colours)
Affixed Reduplication - Prefix I	*jalan* (walk) → *berjalan-jalan* (walking)
Affixed Reduplication - Prefix II	*bukit* (hill) → *berbukit-bukau* (hilly)
Affixed Reduplication - Prefix III	*surat* (letter) → *surat-menyurat* (correspondence)
Affixed Reduplication - Prefix IV	*ajar* (teach) → *pelajar-pelajar*-(students)
Affixed Reduplication - Suffix I	*buah* (fruit) → *buah-buahan* (fruits)
Affixed Reduplication - Suffix II	*makan* (eat) → *makanan-makanan* (foods)
Affixed Reduplication - Confix I	*anak* (child) → *keanak-anakan* (childish)
Affixed Reduplication - Confix II	*tahu* (know) → *pengetahuan-pengetahuan* (knowledge)
Affixed Reduplication - Confix III	*nasihat* (advice) → *nasihat-menasihati* (advice)
Affixed Reduplication - Infix	*jari* (finger) → *jari-jemari* (fingers)
Partial Reduplication	*tamu* (visit) → *tetamu* (visitor)

2.3 Compounding Words and Their Morphological Structures

The compounding words are also another derived words that have two possible morphological structures as described in Table 3.

Table 3. Two possible combinations of compounding words.

Compounding	Combination of two root words with/without affixes
Compounding	Root Word + Root Word *ambil* (to take) and *alih* (to move) → *ambilalih* (to takeover).
Affixed Compounding	Prefix + Root Word + Root Word + Suffix *ambil* (to take) and *alih* (to move) → *pengambilalihan* (merger & acquisition)

The first morphological structure is by combining two root words e.g. *ambil* (to take) and *alih* (to move) to form compounding word, *ambilalih* (to takeover). The second morphological structure is by combining two root words with affixes, usually, confixes e.g. *ambil* (to take) and *alih* (to move) are combined with confixes *peng+an* to form affixed compounding of *pengambilalihan* (merger and acquisition). It is important to note that the root word of affixed compounding words, *pengambilalihan* (merger and acquisition), is not *ambilalih* (to takeover) but *ambil* (to take) and *alih* (to move). Therefore it should not be confused with affixation words whereby root words are identified by removing confixes.

In short, the foundation knowledge on morphological structures of affixation, reduplication and compounding words in Malay language will provide an understanding on the stemming capabilities of the existing word stemmers to remove affixes, clitics, and/or particles. It also provides on how the stemming approaches are selected and also the root cause of stemming errors in the existing word stemmers.

3 Related Works on the Existing Malay Word Stemmers

The first Malay word stemmer was developed for information retrieval application by Othman [16]. Since then, there are many word stemmers have been developed with various stemming approaches [1–3, 7, 8, 11–14, 19, 24, 25]. These word stemmers have been used in various applications such as general purposes [12, 19], information retrieval [1, 2, 7, 8, 14, 16] and text categorization [25]. There are also used in the specific applications such as mobile application for retrieving hadith, query on word translation and query on halal information [4–6, 9, 20–22, 26].

Generally, there are four types of stemming approaches that have been used to develop word stemmers namely affix removal method, successor variety method, dictionary lookup method and n-gram method [23]. However, most of the existing Malay word stemmers only adopted two types of stemming approaches namely affixes removal and dictionary lookup methods due to difficulties in addressing morphological structures of Malay language [1–3, 7, 8, 11, 12, 14, 16, 24]. The combination of affixes removal and dictionary lookup led to the six different stemming approaches in the existing word stemmers namely rule-based word stemming [16], rules application order [2, 11, 12, 24], rules frequency order [1, 7, 8] and modified rule frequency order [14]. Rule-based stemming approach comprises sets of affixes, clitics, and/or particles removal rules that are arranged in alphabetical order and also dictionary lookup to check root word before word stemming [16]. However, it was highlighted that there are overstemming errors using rule-based stemming approach [13]. Therefore the rule application order stemming approach was introduced to improve these stemming errors by rearranging affixes, clitics and particles removal rules in morphological order and the use of dictionary lookup before and after the affixes removal rules in order to check whether the word is a root word [2, 11, 12, 24]. Then, the rule frequency order stemming approach was proposed by considering only most frequent affixes, clitics and particles. Similar to rule application order stemming approach, root word dictionary lookup was also used before and after the affixes, clitics and particles removal rules in order to check whether the word is a root word [1, 7, 8]. Finally, modified the rule frequency order was proposed by using rule frequency order stemming approach but using two different types of dictionaries called background knowledge [14].

Other non-popular stemming approaches proposed by past researchers are modified Porter stemmer [19], Boolean extraction word stemming [25] and syllable-based word stemming [13]. These stemming approaches adopted other techniques to identify and remove affixes, clitics and/or particles from derived words. However, the complexity of morphological structures in Malay language has led past researchers to adopt rule-based stemming approaches to find the correct root words while addressing the existing stemming errors.

4 Word Patterns and Their Respective Stemming Errors

There are four different types of stemming errors in the existing Malay word stemmers namely overstemming, understemming, unstem, and spelling variations and exceptions errors as reported by the past researchers [1–3, 7, 10, 13, 16].

Firstly, overstemming errors occur when word stemmer removes affixes and non-affixes morphemes from derived words e.g. _berkesan_ (effective) → _kesan_ (not _s, kes, san_) where the only affix should be removed is _ber+_ and not _berke+an, ber+an_ or _berke+_. The morpheme _ke_ and _san_ are not affixes but syllables that made up the root word, _kesan_. Secondly, understemming errors occur when word stemmer does not have sufficient rules to stem multiple affixed words during word stemming process e.g. _berkebolehan_ (ability) → _boleh_ (not _keboleh, bolehan, berkeboleh_) whereby there are remaining affixes that must be removed. Thirdly, unstem errors occur when there is no stemming rules for specified affixes e.g. _berpengetahuan_ (acknowledgeable) → tahu (not _berpengetahuan_) whereby there is no stemming rules to remove prefix _berpenge+_ and suffix _+an_. Lastly, special variations and exception errors occur when there is no mechanism to remove prefixes under special conditions whereby the recoding of first letters of the root words are required e.g. _memikir_ (to think) → _fikir_ (not _ikir, mikir_), _memilih_ (to choose) → _pilih_ (not _ilih, milih_).

These stemming errors require word stemmer to recognize word patterns to correctly stem affixation, reduplication and compounding words and also to ensure the order of word stemming rules for selecting and removing the correct affixes from affixation, reduplication and compounding words. Therefore, this paper has reclassified these stemming errors based on their respective root causes during the word stemming process namely affixation stemming errors, reduplication stemming errors and compounding errors. The reclassification of these possible stemming errors provides crucial information for developing an enhanced word stemming rules.

4.1 Affixation Stemming Errors

There are eight different types of possible affixation stemming errors as follows:

Affixation Stemming Errors Type I: This is an overstemming error due to prefixation, suffixation or confixation word stemming rules were applied against the root words that have word pattern similarity with prefixation, suffixation and confixation words e.g. _teruk_ (devastate) → _teruk_ (not _uk_), _taman_ (garden) → _taman_ (not _tam_) and _mentari_ (sun) → _mentari_ (not _tar_).

Affixation Stemming Errors Type II: This is an overstemming error due to confixation word stemming rules were applied against prefixation words that have word pattern similarity with confixation words e.g. *berkawan* (befriend of) → *kawan* (not *kaw*), *dibeli* *(bought by)* → *beli* (not *bel*) and *memakan* (to eat) → *makan* (not *mak*).

Affixation Stemming Errors Type III: This is an overstemming error due to confixation word stemming rules were applied against suffixation words that have word pattern similarity with confixation words e.g. *bersihkan* (to clean up) → *bersih* (not *sih*), *kedainya* (his/her shop) → *kedai* (not *da*) and *petikkan* (quotation) → *petik* (not *tik*).

Affixation Stemming Errors Type IV: There are two possibilities of stemming errors whereby prefixation word stemming rules select incorrect prefixes from prefixation words that lead to whether overstemming error or understemming error: when incorrect prefixes are removed (e.g. *be+* and *ber+*) that lead to overstemming errors e.g. *berasa* (to feel) → *rasa* (not *asa*) and also understemming errors e.g. *beranak* (to give birth) → *anak* (not *ranak*).

Affixation Stemming Errors Type V: There are two possibilities of stemming errors whereby suffixation word stemming rules select incorrect suffixes/clitics/particles from suffixation words that lead to whether overstemming error or understemming error: when incorrect suffixes are removed (*+anlah, +kanlah,* and *+lah*) that lead to overstemming errors e.g. *makanlah* (to eat) → *makan* (not *ma, mak*) and also understemming errors e.g. *biarkanlah* (let it be) → *biar* (not *biark, biarkan*).

Affixation Stemming Errors Type VI: There are two possibilities of stemming errors due to conflicting morphological rules to remove prefixes from prefixation and confixation words and also special variations and exception rules to be added after word stemming process that lead to whether overstemming error, understemming error or special variations and exception error e.g. to select the correct morphological rules of affixes from these words: *mengerikan* (too eerie) → *ngeri* (not *eri, keri, ri*); *mengenal* (to introduce) → *kenal* (*not ngenal, enal, nal*); *mengerat* (to tighten) → *erat* (*not ngerat, kerat, rat*) and *mengetin* (to can up) → *tin* (*not ngetin, ketin, etin*), *menyanyi* (to sing) → *nyanyi* (*not sanyi, yanyi*), *menyapu* (to sweep) → *sapu* (*not nyapu, yapu*).

Affixation Stemming Errors Type VII: This is an understemming error due to multiple affixes in prefixation, suffixation, confixation and infixation words e.g. *pemergianmu* (his/her departure) → *pergi* (not *mergian, mergi*), *terangkanlah* (explain) → *terang* (not *terangk, terangkan*), *kesinambungan* (continuation) → *sambung* (not *sinambung*).

Affixation Stemming Errors Type VIII: This is an unstem error due to no prefixation, suffixation, confixation and infixation stemming rules in word stemming process to remove specified affixes from affixation words e.g. *mengenalkan* (introduce) → *kenal* (*not mengenalkan, ngenalk, ngenal, genal, enal*).

4.2   Reduplication Stemming Errors

There are four different types of possible reduplication stemming errors as follows:

Reduplication Stemming Error Type I: This is an understemming error due to full reduplication stemming rules were applied against affixed reduplication words that have word pattern similarity with full reduplication words e.g. *makanan-makanan (foods)* → *makan* (not *makanan*), *pelajar-pelajar* (students) → *ajar* (not *pelajar*).

Reduplication Stemming Error Type II: This is an overstemming error due to affixed reduplication stemming rules were applied against full reduplication words that have word pattern similarity with affixed reduplication words e.g. *kawan-kawan* (friends) → *kaw* (not *kawan*), *kesan-kesan* (students) → *s* (not *kes*).

Reduplication Stemming Error Type III: There are two possibilities of stemming errors due to conflicting morphological rules of affixes removal selection to remove affixes from affixed reduplication words that lead to whether overstemming error or understemming error e.g. *peralatan-peralatan* (equipments) → *alat* (not *ralat*), *perompak-perompak* (robbers) → *rompak* (not *ompak*).

Reduplication Stemming Error Type IV: This is a spelling variations and exception error in the affixed reduplication word stemming whereby specific affixes requires recoding first letter of root word after stemming process e.g. *penyapu- penyapu* (brooms) → *sapu* (not *nyapu, yapu*), *penolong-penolong* (assistance) → *tolong* (not *nolong*), *pengetua-pengetua* (principals) → *ketua* (not *ngetua*).

Reduplication Stemming Error Type V: This is an unstem error due to no reduplication stemming rules in word stemming process to remove specified affixes from reduplication words e.g. *kawan-kawan* (friends) → *kawan-kawan, penyanyi- penyanyi* (singers) → *penyanyi- penyanyi*.

4.3 Compounding Stemming Errors

There are two different types of possible compounding stemming errors as follows:

Compounding Stemming Error Type I: This is an unstem error due to no compounding stemming rules in word stemming process to reduce compounding words to root words e.g. *ambilalih* (to takeover) → *ambil* (to take) and *alih* (to move) (not *ambilalih*).

Compounding Stemming Error Type II: This is an understemming error due to confixation stemming rules were applied against affixed compounding words that have word pattern similarity with confixation words e.g. *pengambilalihan* (merger & acquisition) → *ambil* (to take) and *alih* (to move) (not *ambilalih*), *ketidaktahuan* (anonymity) → *tidak* (not) and *tahu* (know) (not *tidaktahu*).

In the past research, the existing Malay word stemmers are only focused on addressing affixation stemming errors and still suffer from stemming challenges due to the complexity of morphological structures in Malay language. Therefore, it is highly desirable to develop an enhanced stemming approach that is able to stem affixation, reduplication and compounding words with improved stemming accuracy. The proposed stemming approach must address word stemming rules mismatch to stem affixation, reduplication and compounding words, ensure affixes removal selection to identify and remove correct affixes, clitics and particles from derived words and also develop sufficient word stemming rules based on Malay morphology.

5 The Proposed Word Stemmer

Generally, the proposed word stemmer was developed using Perl Programming v5.5 on MacBook Pro OS X El Capitan. There are seven modules in this proposed word stemmer namely Input module, Word Checking module, Word Pattern Identifier module, Dictionary-based Word Stemming I module, Rule-based Word Stemming module, Dictionary-based Word Stemming II module and Output module. The developments of these modules are based on detailed examination of Malay morphology book [10] and 300 Malay news articles as training datasets from Malaysiakini online news.

The Input module aims to accept text documents, tokenize word patterns by removing special characters and then, normalize the words as standard word patterns. For instance the word *k'jaan* (shorthand for government) will be tokenized as *kjaan* by removing apostrophe (') and then, will be normalized to standard words as *kerajaan* so that word stemming rules will able to stem the word appropriately. Otherwise, word stemming rules will wrongly stem the word *k'jaan* to *k ja* instead of *raja* (king). The tokenization and normalization processes also include word patterns such as email address (*abc@gmail.com*), Internet address (*www.abc.com*), Twitter conversation (*#abcd*), abbreviated root words (*M'sia*) and abbreviated affixation words (*k'tangan-nya*) by using Perl v5.5 regular expression and dictionary lookup method.

The Word Checking module aims to check whether there is a word in the text document for word stemming process or when there is no more word in the text document to be stemmed, to stop word stemming process and then, to trigger the Output module to generate the final output of stemmed words.

Word Pattern Identifier module aims to differentiate root words from derived words by using root word dictionary (5641 words) so that only derived words will undergo word stemming process and at the same time, reducing stemming errors due to word pattern similarity with affixation words (Affixation Stemming Error Type I).

The Dictionary-based Word Stemming I module aims to address stemming errors due to Rule-based Word Stemming module would produce affixation and reduplication stemming errors if it is used alone. With combination of Dictionary-based and Rule-based Word Stemming modules, the affixation stemming errors (Affixation Stemming Errors Type II, III, IV, V, VI, VII) and reduplication stemming errors (Reduplication Stemming Errors Type I, II, III, IV) could be addressed appropriately. There are five types of derivative dictionaries that contain of partial reduplication (39 words), infixation words (41 words), prefixation words (579 words), suffixation words (258 words) and confixation words (1155 words) with their respective root words in Dictionary-based Word Stemming I module.

On other hand, Rule-based Word Stemming module aims to stem affixation and reduplication words whereby there are three types of affixation word stemming rules (57 prefixation word stemming rules, 42 suffixation word stemming rules and 270 confixation word stemming rules) and three types of reduplication word stemming rules (1 full reduplication word stemming rule, 118 rhythmic reduplication word stemming rules and 53 affixed reduplication word stemming rules). These word stemming rules adopts rule application order stemming approaches whereby longest-to-shortest affixes selection is used to identify affixes that to be removed from affixation and reduplication words e.g. *penge+* → *peng+* → *pen+* → *pe+*. Then, the order of prefixation, suffixation, confixation and infixation word stemming rules is arranged by following sequence *confixation* → *prefixation* → *suffixation* due to less stemming errors as reported by Ahmad's word stemmer [2]. Similarly, the order of full, rhythmic and affixed reduplication words is arranged by following sequence *full reduplication* → *rhythmic reduplication* → *affixed reduplication* due to less stemming errors are produced as reported by Kassim's stemmer [2]. Infixation and partial reduplication words are treated in Dictionary-based Word Stemming I module due to these words do not contains any affixes so that dictionary lookup method is the best strategy to stem these words.

The Dictionary-based Word Stemming II module aims to stem compounding words and affixed compounding words to their respective root words while address compounding stemming errors (Compounding Stemming Errors Type I, II). The affixed compounding word will be stemmed using confixation word stemming rules due to it has similar word pattern as confixation words and require extra step to normalize the stemmed word to its root words by using derivative dictionary that contains 119 compounding words with their respective root words e.g. *pengambilalihan* (merger & acquisition) → *ambilalih* (to takeover) → *ambil* (to take) and *alih* (to move).

The last module, Output module, aims to generate an output that contains stemmed words in text document. In short, the proposed word stemmer has a capability to tokenize word patterns, to normalize word patterns and to stem affixation, reduplication and compounding words by using combination of rule-based affixes removal and dictionaries lookup methods. This stemming approach is called enhanced rule application order whereby the word stemming rules are arranged in the following sequence dictionary-based word stemming rules (*partial reduplication* → *infixation* → *confixation* → *prefixation* → *suffixation*) → rule-based word stemming rules (*confixation* → *prefixation* → *suffixation* → *full reduplication* → *rhythmic reduplication* → *affixed reduplication*). This enhanced rule application order stemming approach is able to address possible affixation stemming errors (Affixation Stemming Errors Type I–VIII), reduplication stemming errors (Reduplication Stemming Errors Type I–V) and compounding stemming errors (Compounding Stemming Errors Type I–II). The complexity of Malay language leads to the use of combined stemming approaches particularly affixes removal and dictionary lookup methods whereby a single stemming approach would not able to address all possible stemming errors such as root word that are similar to derived word patterns, multiple morphological rules and conflicting affixes removal selection from derived words. The proposed word stemmer is described in the following pseudo codes:

```
 1 Input: Input Module (Processing the Input)
 2 Accept the input text document.
 3 Remove special characters except hyphen, quotes
 4 Normalize the word
 5 Go to Step-1.
 6
 7 Output:Output Module (Processing the Output)
 8 Root Word {stem1,stem2...stemn})
 9
10 Step-1: Word Checking Module (Check Words in Document)
11 IF i=0, go to Output.
12 IF i=wordn, go to Step-2.
13
14 Step-2: Word Pattern Identifier Module
15 (Differentiate Root Words from Derived Words)
16 IF i=wordn is root word, accept word, go to Step-1.
17 ELSE i=wordn is derived word, go to Step-3.
18
19 Step-3: Dictionary-based Word Stemming I Module
20 (Word Stemming Using Dictionary Lookup)
21 IF i=partial reduplication, stem word, go to Step-1.
22 IF i=infixation, stem word, go to Step-1.
23 IF i=confixation, stem word, go to Step-1.
24 IF i=prefixation, stem word, go to Step-1.
25 IF i=suffixation, stem word, go to Step-1.
26 ELSE go to Step-1.
27
28 Step-4: Rule-based Word Stemming Module
29 (Word Stemming Using Affixes Removal Method)
30 Condition I - Word Without Hyphen (-)
31 IF i=confixation, stem word, go to Step-5.
32 IF i=prefixation, stem word, go to Step-5.
33 IF i=suffixation, stem word, go to Step-5.
34 ELSE accept word, go to Step-5.
35 Condition II - Word with Hyphen (-)
36 IF i=full reduplication, stem word, go to Step-5.
37 IF i=rhythmic reduplication, stem word, go to Step-5.
38 IF i=affixed reduplication, stem word, go to Step-5.
39 ELSE accept word and go to Step-5.
40
41 Step-5: Dictionary-based Word Stemming II Module
42 (Compounding Word Stemming Using Dictionary Lookup)
43 IF i=compounding, stem word, go to Step-1.
44 ELSE accept word, go to Step-1.
```

6 Experimental Results and Discussion

In order to evaluate the performance of the proposed word stemmer, 500 Malay news articles from Malaysiakini news agency have been used as testing dataset that consist of 112,785 word occurrences with 20,853 unique words. The result of the experiment is described in Table 4.

Table 4. Experimental results of the proposed affixation word stemmer.

Experiment	Stemming accuracy	Stemming errors examples
Proposed word stemmer against 500 news articles	88.5 %	1. Stemming Errors at Input Module: *pro-kerajaan, menjadiseperti, me-laksanakan* 2. Stemming Errors at Rule-based Word Stemming Module: *tribahasa, muslimat, subtajuk, kilometer, hadirin, poliklinik* 3. Stemming Errors at Dictionary-based Word Stemming II Module: *ketidakperimanusiaan, ketidakperikemanusian* 4. Others: name (*nani, nazri, osman*), place (*jerman, kalimantan, kedah*), brands (*jpmorgan, perodua, proton*), English words (*indian, malaysian, product*)

The experimental result indicates that the proposed word stemmer achieved 88.5 % stemming accuracy to stem affixation, reduplication and compounding words. It can be concluded that this proposed word stemmer performs better than the existing word stemmers due to there is no affixation stemming errors occurred during word stemming against testing datasets. However, the proposed word stemmer faces new stemming errors with the following categories: (a) unable to tokenize affixation words with hyphen (*pro-kerajaan*), combination of two words (*menjadiseperti*), word at end-of-line (*me-laksanakan*). (b) unable to remove loaned and scientific affixes e.g. *tri +*, *sub+* and *+in* whereby only standard affixes are considered. (c) unable to stem complex affixed compounding words (*ketidakperimanusiaan, ketidakperikemanusiaan*) and (d) unable to differentiate proper nouns words that have word pattern similarity to affixation words e.g. person's name (*osman*), place (*jerman*), brands (*proton*) and English word (*product*). These stemming errors do not fall into the possible word stemming errors as discussed in Sect. 4 and therefore, are not considered during development of the proposed word stemmer. None of these stemming errors has ever been discussed in the past research. In order to improve these stemming errors, further enhancement of the proposed word stemmer must be made in our future works as follows: (a) to tokenize affixation words with hyphen (*pro-kerajaan, al-kitab, sub-kontraktor*), word at end-of-line (*me-laksanakan, melak-sanakan, melaksa-kan*) and also combined two root words (*menjadiseperti, akanbegitu, laludia*) at Input module (b) to stem affixation words with loaned and scientific affixes such as (Sanskrit affixes - *dasa+, eka+, +man*, English affixes - *pro+, sub+*, Arabic affixes - *+ah, +in, +at*, Scientific affixes – *kilo+, mega+, tera+)* at Rule-based Word Stemming module (c) to add complex affixed compounding words in derivative dictionary entries at Dictionary-based Word Stemming II module and (d) to add dictionary entries that comprises proper nouns that may affect word stemming at Word Pattern Identifier module. In short, the proposed word stemmer is able to stem affixation, reduplication and compounding words with promising stemming accuracy.

7 Conclusion

This paper describes an enhanced word stemmer for Malay language that uses four different word treatment processes namely tokenization and normalization, dictionary lookup stemming approach, ruled-based stemming approach and compounding word normalization. Unlike the existing Malay word stemmers, the proposed word stemmer is able to stem affixation, reduplication and compounding words and addresses three different types of stemming errors namely affixation stemming errors, reduplication stemming errors and compounding stemming errors. To address these stemming errors, rule-based word stemming approach for straightforward morphological rules and dictionary-based word stemming approach for conflicting or multiple morphological rules were developed in order to identify the correct affixes, clitics, and/or particles from affixation, reduplication, and compounding words. Our future work will focus on developing word spell checker, adding loaned and scientific affixes in the word stemming rules and dictionary entries for proper nouns such as person's name and place.

Acknowledgments. The authors would like to thank the Editor-in-Chief and the anonymous reviewers of the manuscript for their valuable comments and suggestions. This research was funded by Universiti Teknologi Malaysia's Research University Grant PY/2014/02479.

References

1. Abdullah, M.T., Ahmad, F., Mahmod, R., Sembok, T.M.T.: Rules frequency order stemmer for Malay language. IJCSNS Int. J. Comput. Sci. Netw. Secur. **9**(2), 433–438 (2009)
2. Ahmad, F., Yusoff, M., Sembok, T.M.: Experiments with a stemming algorithm for malay words. J. Am. Soc. Inform. Sci. **47**(12), 909–918 (1996)
3. Alfred, R., Leong, L.C., On, C.K., Anthony, P.: A literature review and discussion of malay rule-based affix elimination algorithms. In: 8th International Conference on Knowledge Management in Organizations. Springer Proceedings in Complexity, pp. 285–297. Springer, Netherlands (2014)
4. Al-Ramahi, M., Mustafa, S.: N-Gram-Based Techniques for Arabic Text Document Matching, Case Study: Courses Accreditation (2012)
5. Al-Shalabi, R., Kannan, G., Hilat, I., Ababneh, A., Al-Zubi, A.: Experiments with the successor variety algorithm using the cutoff and entropy methods. Inf. Technol. J. **4**(1), 55–62 (2005)
6. Bakar, Z.A., Rahman, N.A.: Evaluating the effectiveness of thesaurus and stemming methods in retrieving malay translated Al-Quran documents. In: Sembok, T.M.T., Zaman, H.B., Chen, H., Urs, S.R., Myaeng, S.-H. (eds.) ICADL 2003. LNCS, vol. 2911, pp. 653–662. Springer, Heidelberg (2003)

7. Darwis, S.A., Abdullah, R., Idris, N.: Exhaustive affix stripping and a malay word register to solve stemming errors and ambiguity problem in malay stemmers. Malays. J. Comput. Sci. **25**(4), 196–209 (2012)
8. Fadzli, S.A., Norsalehen, A.K., Syarilla, I.A., Hasni, H., Dhalila, M.S.S.: Simple rules malay stemmer. In: The International Conference on Informatics and Applications (ICIA 2012). The Society of Digital Information and Wireless Communication, pp. 28–35 (2012)
9. Hanum, H.M., Bakar, Z.A., Rahman, N.A., Rosli, M.M., Musa, N.: Using topic analysis for querying halal information on malay documents. Procedia Soc. Behav. Sci. **121**, 214–222 (2014)
10. Hassan, A.: Morfologi. PTS Professional, vol. 13 (2006)
11. Idris, N., Syed, S.M.F.D.: Stemming for term conflation in malay texts. In: International Conference on Artificial Intelligence (ICAI 2001) (2001)
12. Kassim, M.N., Maarof, M.A., Zainal, A.: Enhanced rules application order approach to stem reduplication words in malay texts. In: Herawan, T., Ghazali, R., Deris, M.M. (eds.) Recent Advances on Soft Computing and Data Mining SCDM 2014. AISC, vol. 287, pp. 657–665. Springer, Heidelberg (2014)
13. Lee, J., Othman, R.M., Mohamad, N.Z.: Syllable-based Malay word stemmer. In: 2013 IEEE Symposium on Computers and Informatics (ISCI), pp. 7–11 (2013)
14. Leong, L.C., Basri, S., Alfred, R.: Enhancing malay stemming algorithm with background knowledge. In: Anthony, P., Ishizuka, M., Lukose, D. (eds.) PRICAI 2012. LNCS, vol. 7458, pp. 753–758. Springer, Heidelberg (2012)
15. Lovins, J.B.: Development of a stemming algorithm, MIT Information Processing Group, Electronic Systems Laboratory (1968)
16. Othman, A.: Pengakar Perkataan Melayu untuk Sistem Capaian Dokumen, MSc Thesis, Universiti Kebangsaan Malaysia, Bangi, Malaysia (1993)
17. Porter, M.F.: An algorithm for suffix stripping. Program **14**(3), 130–137 (1980)
18. Ranaivo-Malancon, B.: Computational analysis of affixed words in malay language. In: Proceedings of the 8th International Symposium on Malay/Indonesian Linguistics, Penang, Malaysia (2004)
19. Sankupellay, M., Valliappan, S.: Malay language stemmer. Sunway Acad. J. **3**, 147–153 (2006)
20. Sembok, T.M.T., Yussoff, M., Ahmad, F.: A malay stemming algorithm for information retrieval. In: Proceedings of the 4th International Conference and Exhibition on Multi-lingual Computing, vol. 5, pp. 1–2 (1994)
21. Sembok, T.M., Willett, P.: Experiments with n-gram string-similarity measure on Malay texts, Universiti Kebangsaan Malaysia. Bangi, Malaysia **22**, 335–345 (1995)
22. Sembok, T.M.T., Bakar, Z.A.: Effectiveness of stemming and n-grams string similarity matching on malay documents. Int. J. Appl. Math. Inform. **5**(3), 208–215 (2011)
23. Sharma, D.: Stemming algorithms: a comparative study and their analysis. Int. J. Appl. Inf. Syst. **4**(3), 7–12 (2012)
24. Tai, S.Y., Ong, C.S., Abdullah, N.A.: On designing an automated malaysian stemmer for the malay language. In: Proceedings of the Fifth International Workshop on Information Retrieval with Asian Languages, pp. 207–208. ACM (2000)
25. Yasukawa, M., Lim, H.T., Yokoo, H.: Stemming malay text and its application in automatic text categorization. IEICE Trans. Inf. Syst. **92**(12), 2351–2359 (2009)
26. Zainudin, M.K.A.B., Rias, R.M.: M-Hadith: retrieving malay hadith text in a mobile application. In: 2012 IEEE Symposium on Computer Applications and Industrial Electronics (ISCAIE), pp. 60–63 (2012)

Building a Process Description Repository with Knowledge Acquisition

Diyin Zhou, Hye-Young Paik[✉], Seung Hwan Ryu, John Shepherd,
and Paul Compton

School of Computer Science and Engineering,
University of New South Wales, Sydney, Australia
{dzho186,hpaik,seungr,jas,compton}@cse.unsw.edu.au

Abstract. Although there is an abundance of how-to guides online, systematically utilising the collective knowledge represented in such guides has been limited. This is primarily due to how-to guides (effectively, informal process descriptions) being expressed in natural language, which complicates the process of extracting actions and data. This paper describes the use of Ripple-Down Rules (RDR) over the Stanford NLP toolkit to improve the extraction of actions and data from process descriptions in text documents. Using RDR, we can incrementally and rapidly build rules to refine the performance of the underlying extraction system. Although RDR has been widely applied, it has not so far been used with NLP phrase structure representations. We show, through implementation and evaluation, how the use of action-data extraction rules and knowledge acquisition in RDR is both feasible and effective.

1 Introduction

In daily life, people often undertake processes[1] whose intrinsic details are unknown because they are being encountered for the first time. Such processes could be as simple as cooking a dish or as complex as buying a house. To distinguish them from the conventional (and more widely-studied) organisational workflows, we refer to these kinds of processes as *"personal processes"*. Descriptions of personal processes are often shared on-line in the form of how-to guides (e.g., on-line recipe sites, eHow[2], WikiHow[3]), which are normally written in natural language as a set of step-by-step instructions.

Although the collective knowledge and data represented in such descriptions could be useful in many applications, the potential for using them is curbed by their natural language format. The ultimate goals of our project [3] are (i) to build a repository of personal process descriptions based on a formal model and (ii) to investigate novel query techniques and data analytics on such descriptions.

---

[1] In this paper, we use the terms process and workflow interchangeably.
[2] www.ehow.com, eHow.
[3] www.wikihow.com, WikiHow.

© Springer International Publishing Switzerland 2016
H. Ohwada and K. Yoshida (Eds.): PKAW 2016, LNAI 9806, pp. 86–101, 2016.
DOI: 10.1007/978-3-319-42706-5_7

In this paper, as a step towards building the repository, we describe *RDR-ADE* (RDR Action Data Extractor), which aims to extract the basic constructs of a process description from how-to guides.

We adopt a knowledge acquisition approach (using Ripple-Down Rules [1]) to allow human users to improve the performance of a standard NLP parser [10] in identifying verbs and objects in personal process descriptions. Although RDR has been used in conjunction with NLP tools in the past, the aim of our approach is to extract basic process constructs (i.e., actions and data) from text, which has not been done with RDR before. Specifically, we make the following contributions.

- We propose the notion of *action-data extraction rules* (henceforth *extraction rules*). These rules represent user knowledge about how the actions and their associated data can be identified from process descriptions.
- We present *incremental knowledge acquisition techniques* to build and update a set of extraction rules. The continuous update of extraction rules enables our system, over time, to make finer-grained extraction decisions.
- We present an implementation of *RDR-ADE* and provide the evaluation results that show the feasibility and effectiveness of our proposed approach.

While the ideas we present should apply to many types of process descriptions, we consider here only cooking recipes due to the availability of large numbers of relatively compact online descriptions[4].

The rest of the paper is structured as follows: In Sect. 2, we discuss the use of RDR in other systems as a way to improve the underlying techniques. We then briefly introduce some background concepts such as a model for personal processes and Stanford CoreNLP. In Sects. 4, 5 and 6 we describe the details of the rule models and implementation, followed by evaluation results and conclusion.

2 Related Work

Ripple Down Rules (RDR) [1] is a knowledge acquisition technique that allows experts to rapidly construct knowledge bases on a case-by-case basis while a system is already in use. It can be built incrementally on top of an existing system to improve the underlying system. RDR and its variants have been successfully used in a wide range of application domains, such as Web document classification [8], diagnosis [9], information extraction [7], and have had significant commercial uptake [2,9]. A wide range of RDR research covering a range of RDR methods is reviewed in [12].

The key idea behind RDR is that cases are processed by the RDR system and when the output provided by the system is not correct, a new rule is added to give the correct output for that case. In this process, the expert does not need to consider the structure of the knowledge base or modify other rules. They build the rule to give the correct conclusion for the case they are trying to correct,

---

[4] Recipes are also useful for subsequent studies in personal process analysis, but that is beyond the scope of this paper.

and in most RDR systems this rule is then checked against cases for which rules were previously added to see that none of these cases are misclassified. If necessary the expert adds further rule conditions, making the rule more specific for the case in hand excluding the previous cases. The system then automatically adds the new rule into the knowledge base. This is a very rapid process and log data from commercial systems shows that across 10 s of 1000 s of rule and 100 s of knowledge bases, the average time to add a rule was under two minutes [9].

In terms of utilising RDR with NLP tools, [6] shows how the informal writing style of Web documents tends to negatively affect the performance of NER (Named Entity Recogniser) parsers. The paper proposes an RDR-based knowledge acquisition process where the rules are used to correct spelling errors or missing/unclassified named entity (NE) tags (e.g., 'YouTube' and 'YOUTUBE' should be classified as 'ORG' (organisation)). The same authors also applied the same principles to extract *Open Relations* between named entities. Open Relation Extraction systems seek to extract all potential relations from the text rather than extracting pre-defined ones [7]. The only previous work on RDR for parts of speech and phrase representations was an automated learner using the RDR structure, rather than actual knowledge acquisition [11].

Our aim is to not only demonstrate improvement on the NLP Stanford parser, but to improve the extraction of action and data information from the text, which is something that has not been done before with RDR.

3   Preliminaries

In this section, we give the background concepts that are necessary to describe the *RDR-ADE* system: the target schema that *RDR-ADE* works towards and the extraction process of action/data pairs using the Stanford NLP parser.

3.1   PPDG: A Model for Personal Process Descriptions

To formally represent and analyse the process instructions in how-to guides, we proposed a graph-based model, called "Personal Process Description Graph" and a query language over this model [5,14]. PPDGs are labelled directed graphs that include both actions and data. In a PPDG, there are different node types for actions and data, different arc types to represent action flow and data flow, and associations between actions and data. The formal definitions and properties of the model are omitted here as the main topic of this paper does not directly involve PPDGs. Our goal is to extract, from a process description, the information required to define action nodes and data nodes to build a PPDG.

3.2   Action and Data Extraction Process

The Stanford CoreNLP toolkit [10] provides a collection of NLP tools. The tools that are important for our purposes are the part-of-speech (POS) tagger, which

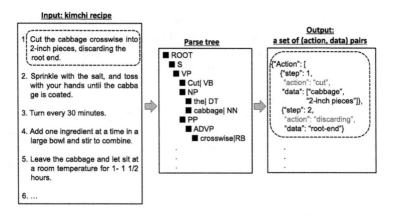

Fig. 1. Example input and output for the *RDR-ADE* system

marks words with their syntactic function, and the parser, which produces a parse tree for a chunk of natural language text.

The overall process of extracting actions and data from a description is shown in Fig. 1. We break this process into the following phases:

1. *Input Text Segmentation.* A process description consists of a number of "input units" where each "input unit" contains one or more sentences and each sentence contains one or more phrases. Each phrase corresponds to a single step in the process. Input units are separated by line breaks. The parser partitions each input unit into sentences and phrases.
2. *Parse Tree Generation.* Using the Standford NLP parser, this phase takes an input unit and produces as output a parse tree. As in Fig. 1, the non-leaf nodes of the parse tree carry the phrase structure (via tags such as NP (Noun Phrase) or VP (Verb Phrase)), while leaf nodes carry the part-of-speech (POS) for individual words (via tags such as NN (Noun), VB (Verb)).
3. *Action and Data Extraction.* In this phase, we extract a set of (action, data) pairs from the parse trees generated from the previous phase. In identifying each pair, we use the phrase structure and consider a word with VB (and derivations of it) as an action and NN as a data item. The extracted pairs are written as JSON objects for further processing (e.g., mapping to a PPDG).

The phrase structure contained in a parse tree is useful for identifying the most likely places from which to find verbs and objects. In each sentence, the parse tree tags a verb phrase (VP) and within a verb phrase, we may find a relevant noun phrase (NP).

While the NLP toolkit is useful for our purpose, it has limitations for the kind of informal text that we typically find in process descriptions.

For example, the default POS tagger and parser are trained to assume that the first phrase in a sentence will be a noun phrase referring to the sentence subject. However, recipe sentences generally start with the verb, and have an implicit subject ("you"). This results in common types of recipe sentences (e.g. "smoke the salmon") being misinterpreted as starting with a noun (e.g. "smoke") if the first word can be interpreted as a noun. Another type of sentence which caused problems for the parser was one in which an adverbial phrase was placed at the start of the sentence (e.g. "with a sharp knife, chop the cabbage" vs "chop the cabbage with a sharp knife").

As we have briefly outlined in Sect. 2, instead of re-training the CoreNLP tools, we employ a similar approach to [6,7]. We exploit the RDR technique to build and update action-data extraction rules over the baseline system (i.e., CoreNLP tools). These rules are designed to address the said problems. The continuous updates of extraction rules enable our system to perform more precise extraction.

4   Harnessing User Knowledge

In this section, we describe the notion of *extraction rules* and their management. As noted above, previous RDR techniques on NLP tools were targeted on extracting named entities or open relations - utilising POS tags. The knowledge we need to represent in *RDR-ADE* has to be expressed over the parse tree structure. The rule model we present below is designed to express user knowledge about which situations the actions and their relevant data can be identified in a given parse tree.

4.1   Extraction Rule Representation Model

Each rule has two components: a condition and a conclusion. The conclusion part of the rule simply states how a word is to be labelled as 'action' or 'data', or left unlabelled. A condition consists of a conjunction of predicates on parse trees. To express a condition relating to the parse tree, we propose the following rule syntax components: nodes, test values and operators.

Nodes: To express conditions over nodes in the parse tree, we provide intuitive access names for the nodes (following the XML document model). Some of the examples of the possible names are: `currentNode`, `parentNode`, `allAncestors`, x^{th}`LevelParentNode`, `firstChild`, `lastChild`, `nextSibling`, `prevSibling`, etc.

Test Values: The test values can be of two types: Tags or Regular Expressions. Tags represent the parse tree tag values that a node could be associated with,

Table 1. A sample list of tags

PT*	Description	WT+	Description	WT+	Description
ADVP	Adverb phrase	CD	Cardinal number	NN, NNS	Noun, plural
NP	Noun phrase	DT	Determiner	VB, VBG	Verb, gerund
PP	Prepositional phrase	IN	Preposition	VBN	Verb past participle
VP	Verb phrase	JJ	Adjective	PRP$	Possessive pronoun

*PT = Phrase Level Tags, +WT = POS Word Tags.

such as VP, NN, and DT. Table 1 describes some of the phrase structure tags and part-of-speech tags[5] used in our system. Besides the standard tags, we have our own custom tags: 'ACTION' and 'DATA'.

A test value could also contain a regular expression. This is useful, for example, when the user wants to match any POS tags that are derivations of a base form (e.g., VBD, VBG, VBN, VBP, VBZ are derivations of VB). Note that a regular expression could also include a literal value (e.g., 'Oil').

Operators: The set of operators we currently support allows for a given node to be tested against certain properties. For example, we could test if a node is a verb phrase, or has a text value of 'X'. The design and implementation of these operators is at the heart of the rule design. Our current implementation supports the following operations[6]:

- `HasPhraseTag`: returns true if the node tested has a phrase tag give in the test value, e.g., PP (Prepositional phrase), VP (Verb phrase).
- `HasWordTag`: returns true if the node tested has a part-of-speech tag give in the test value, e.g., DT (Determiner), CD (Cardinal number), VB (Verb). We use the term word tags for part-of-speech tags.
- `HasActionObjectTag`: returns true if the node tested is labelled with the our custom tags "Action" or "Data".
- `IsLeafNode`: returns true if the node tested is a leaf node in the parse tree.
- `HasText`: returns true if the node tested has the text given in the test value.
- `CanBeOfWordType`: returns true if the text value of the node tested is of the word type given in the test value, e.g., (NN) Noun, (VB) Verb, or (JJ) Adjective.

We use the WordNet API[7] to implement the `CanBeOfWordType()` operator. This operator is used to see if a given word could have different functions in a sentence. For example, 'oil' could be a noun (as in 'olive oil') or a verb (as in 'oil the fish').

Using these components, we define an extraction rule as follows:

Definition 1 (Extraction Rule). *Let N_t be a set of nodes in a parse tree t. An extraction rule has the form:* IF P THEN C *where P is a conjunction of predicates on tree nodes and test values, and C is a conclusion. Each predicate has the form* $(node, op, val)$, *in which* $node \in N_t$, $op \in \{HasPhraseTag, IsLeafNode, ...\}$, *and* $val \in \{VP, NN, VB, NP, ...\}$. *The conclusion has the form* $(node, action/data)$.

For example, the rule (`currentNode,HasWordTag,NN`)→(`currentNode,`'Data') checks whether the current tree node has a tag of NN and then determines that it must be a data item. If the node was `cabbage|NN`, the conclusion would be that "cabbage" was data for the current action.

---

[5] For the complete set of part-of-speech tags generated by the Standford parser, see http://www.comp.leeds.ac.uk/amalgam/tagsets/upenn.html.

[6] However, adding a new operator is a straightforward task in our system.

[7] WordNet 3.0, https://wordnet.princeton.edu.

4.2 Matching Extraction Rules

When a parse tree t is generated from a case, the system identifies two extraction rules where one is for identifying actions and the other is for identifying their associated data. For this, we provide an operation called `MatchRule()`, which takes as input t and produces as output the rules for the action and data identification process. The system matches t against the conditions of a set of extraction rules. It evaluates the rules at the first level of rule tree. Then, the next level of rules are evaluated, if their parent rules are satisfied. If no extraction rule is found to be appropriate to t, the user might build a new rule with the help of rule editor provided by our system.

Algorithm 1. MatchRule

Input: Parse tree t and a set of extraction rules R
Output: A set of matched extraction rules
begin
1:    Let $satisfiedRules := \phi$;
2:    // C is a condition of a rule
3:    // p is a predicate of C
4:    **foreach** $r \in R$ **do**
5:        $C := getCondition(r)$;
6:        $allPredicatesSatisfied :=$ true;
7:        **foreach** $p \in C$ **do**
8:            **if** not $isSatisfiedBy(p, t)$ **then**
9:                $allPredicatesSatisfied := false$;
11:            **endif**
11:        **endfor**
12:        **if** $allPredicatesSatisfied$ **then**
13:            $satisfiedRules := satisfiedRules \cup r$;
14:        **endif**
15:    **endfor**
16:    **return** $satisfiedRules$;
end

---

5   Incremental Knowledge Acquisition

This section presents how to incrementally obtain the extraction rules from users.

5.1 Knowledge Acquisition Method: Ripple Down Rules

To build and update extraction rules, we use the RDR [1] knowledge acquisition method because: (i) it provides a *simple* approach to knowledge acquisition and maintenance; (ii) it works *incrementally*, in that users can start with an empty rule base and gradually add rules while processing new cases.

RDR organizes the extraction rules as a tree. In *RDR-ADE*, we have two rule trees: one for action extraction (Fig. 2(a)), the other for data extraction (Fig. 2(b)). For example, the rule tree for action extraction has *Action_DefaultRule* and *Action_BaseRule*. The exceptions to the base rule are named *AE_Rule1*,

AE_Rule2, ... AE_RuleX according to the creation order. *Action_DefaultRule* is the rule that is fired initially for every *case*. The rules underneath it are more specialized rules created by adding exception conditions to their parent rules. The *rule inference* in RDR starts from the root node and traverses the tree, until there are no more children to evaluate. The conditions of nodes are examined via depth-first traversal, which means the traversal result is the rule whose condition is satisfied *last*. The same applies to the rule tree for data extraction. We note that for each case, *RDR-ADE* evaluates both the action extraction and data extraction rule trees to produce the JSON output.

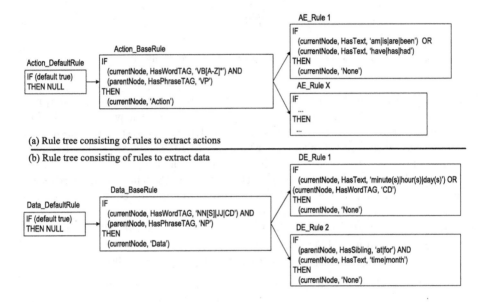

Fig. 2. Example RDR trees (abbreviated)

5.2   Acquiring User Knowledge Incrementally

In what follows, we demonstrate how error-correcting rules are acquired from the user incrementally using a sequence of cases as an example scenario. In the cases, actions are underlined and data is shown in **bold**.

Case 1. "Cut the **cabbage** crosswise into **2-inch pieces**, discarding the **root end**".

From the sentence in Case 1, our system generates the parse tree shown in Fig. 3. At this point, there is one default rule in each rule tree. These rules are applied to this parse tree and NULL values are returned from each rule tree.

The user considers this as an incorrect result and adds new rules *Action_BaseRule* and *Data_BaseRule* under the default rules as shown in Fig. 2.

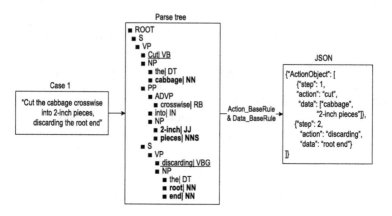

Fig. 3. Case 1 applying two new exception rules

Action_BaseRule specifies that, if a node has a word tag matching a regular expression 'VB[A-Z]*' and its parent node has a phrase tag 'VP', the word is labelled as 'Action'. By applying this rule to the parse tree of Case 1, the system returns a set of actions {cut, discarding}. On the other hand, *Data_BaseRule* states that if a node has a word tag 'NN[S]', 'JJ' or 'CD' and its parent node has a phrase tag 'NP', the word should be labelled as 'Data'. From the parse tree, this rule returns {cabbage, 2-inch pieces, root end}. Figure 3 shows the results of applying these two new rules to the case, which is now considered correct.

In fact, as indicated by their names, we consider these two rules as the base rules in our system for extracting actions and data respectively.

Now we consider the next case, Case 2 whose parse tree is shown in Fig. 4.

Case 2. "Sprinkle with **salt**; toss with your **hands** until the cabbage is coated".

Using the parse tree for this case, the two base rules are fired and the system returns as actions {sprinkle, toss, is coated} and as data {salt, hands, cabbage}.

The user considers the results and decides to exclude 'is coated' from the action list. As a general rule in our system, we ignore forms of the verb 'to be' (and sometimes 'to have') when used as an auxiliary together with the past participle of a transitive verb, especially when a word like 'until' is used as a subordinating conjunction to connect another action to a point in time.

To ignore BE-verbs (e.g. am, are, is, been, ...) and HAVE-verbs (have, has, had, ...) from the actions, the user adds the following rule as an exception to *Action_BaseRule*:

AE_Rule1: For current node $n \in N_t$, (n, HasText, 'am,is,are,been') or (n, HasText, 'have,has,had') → *n is not labelled as 'action'.*

According to *AE_Rule1*, if the current node contains either a BE-word or a HAVE-word, the word associated with the node is ignored from action labelling. Thus, from the same parse tree in Fig. 4, the rule matching algorithm now generates the final JSON object by applying *AE_Rule1* and *Data_BaseRule* instead

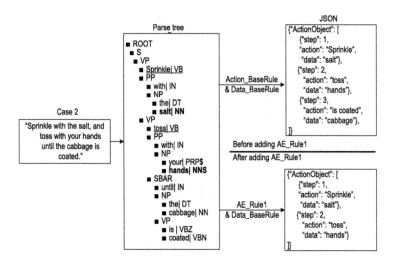

Fig. 4. Case 2 ignoring HAVE-verbs and BE-verbs.

of *Action_BaseRule* and *Data_BaseRule*. Here, we do not extract the cabbage as data because its associated verb is not identified as an action.

Case 3. "<u>Turn</u> every 30 minutes".

In Case 3, according to the existing rules so far, 30 minutes is classified as 'Data' (by *Data_BaseRule*). In this scenario, the user decides to ignore numbers or units such as 30 minutes, 2 days, 30 cm and so on, because she considers them as auxiliary information that is certainly useful but not part of the key action/data constructs in a process. She may want to consult the system developer to define a new type of label "UNITS", but for now, she adds a new rule *DE_Rule1* as an exception of *Data_BaseRule*.

> *DE_Rule1: For the current node $n \in N_t$, (n, HasText, 'minutes(s), hour(s), day(s)') or (n, HasWordTag, 'CD') → n is not labelled as 'data'.*

DE_Rule1 states that, if a node has a time-word or has a word tag CD or DT, then it is not labelled as data. After this rule is defined, in the final JSON object in Fig. 5(a), we extract only the action "turn" from the sentence.

Case 4. "<u>Add</u> **one ingredient** at a time in a **large bowl** and <u>stir</u> to <u>combine</u>".

Now consider Case 4. According to the rules so far, *Data_BaseRule* will make time|NN, under NP as 'Data'. The user considers that the result is not what she expected, and decides that she does not want to extract data from propositional phrases such as at a time, in half, for up to one month, etc. She adds the following rule *DE_Rule2*.

> *DE_Rule2: For current node $n \in N_t$, parent node $pn \in N_t$, (pn, HasSibling, 'at,for') and (n, HasText, 'time,month') → n is not labelled as 'data'.*

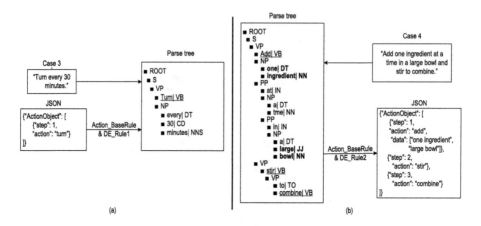

Fig. 5. Case 3 ignoring numbers or units, Case 4 ignoring prepositional phrases.

The rule says if a current node is a time-related word and its parent node has a sibling node tagged as at or for, then word associated with the node is not labelled 'Data'. The final JSON objects with this rule is shown in Fig. 5(b).

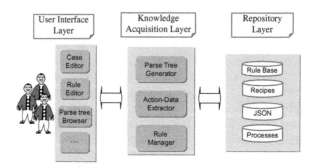

Fig. 6. *RDR-ADE* system architecture

6 Implementation and Evaluation

This section describes our prototype implementation and experimental results.

6.1 Implementation

A prototype was implemented using Java, J2EE and Web services. The RDR engine and action-data extractor are all independent Java programs that are wrapped by a REST web service and accessed through HTTP. This architecture allows other web pages or applications to make use of the services in other ways. The *RDR-ADE* system consists of the following three layers: user interface, knowledge acquisition, and repository (see Fig. 6).

The *user interface layer* allows users to browse generated parse trees and incrementally build extraction rules using the rule editor. The *knowledge acquisition layer* is responsible for generating parse trees, extracting actions and data, creating rules, etc. The *repository layer* stores the rules, process descriptions (e.g., recipes), JSON objects, and so on. Table 2 shows a set of operations that the components of such layers can invoke to carry out their specific functions. Figure 7 gives a screen-shot of our system. Here, we see the input case on the top left panel. For the input case, the system generates a parse tree in the bottom right panel. Then, using the extraction rules in a knowledge base, it produces a set of (action, data) pairs in the format of a JSON object in the top right panel.

Table 2. The list of operations invoked in *RDR-ADE*

Parse tree generator/Rule manager operations
- *generateParseTree(c)* produces a parse tree from an input case *c*.
- *matchRule(c)* returns a list of extraction rules applicable to an input case *c*.
- *createRule(c,d)* creates a rule with a condition *c* and a conclusion *d*.
- *refineRule(r,c,d)* refines a rule *r* with a condition *c* and a conclusion *d*.
Action and data extractor operations
- *extractActionData(p)* identifies actions and data from a given parse tree *p*.
- *generateJSON(ad)* generates a JSON object from a set of (action, data) pairs *ad*.

6.2 Evaluation

We now present the evaluation results to show how effectively the *RDR-ADE* system identifies actions and their associated data from process descriptions.

Dataset. We use a dataset derived from 30 recipes. The dataset consists of 317 sentences and 4765 words. We have manually labelled the verbs (as 'action') and data items (as 'data') in each sentence to create the ground-truth. Each sentence is uniquely identified with an ID. We then processed sentences one by one in the presented order.

Evaluation Metrics. We measured the overall performance of the extraction system using the following formula.

$$\text{Accuracy} = \frac{\text{the number of correctly identified actions and data items}}{\text{the total number of labelled actions and data items}}$$

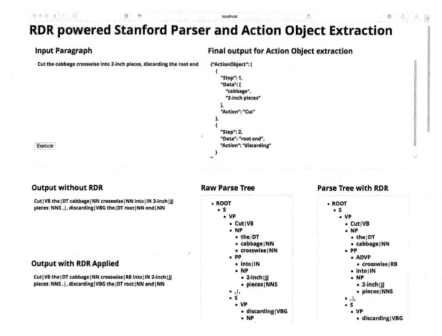

Fig. 7. Screenshot showing: input case, parse tree, and action-data pairs.

Training Phase. Starting with an empty knowledge base, we began the knowledge acquisition process by looking at the sentences one by one in the order prepared at the start of the experiment.

The acquisition process is defined as follows (note that this process is repeated for every sentence): (i) a sentence is given as an input case, (ii) rules are applied, (iii) we examine the result, (iv) if the result is what the user expected, the rule-base is untouched; if not, an exception rule is added to the rule base.

The above steps are repeated until all sentences are considered, or until we do not see significant improvement in the accuracy measure. With an RDR system one can keep adding rules indefinitely as increasingly rare errors are identified. In critical in application areas such as medicine, the ability to keep easily adding rules if required is a key advantage of RDR. In other domains, and in research studies such as this, it is sufficient to add rules until the performance plateaus and adding new rules has a negligible effect on the overall performance.

In the first run, we stopped at the 212th case. In the following discussion we have called the initial 212 cases "training data" and used the remaining cases (cases 213-317) as the "test data" (i.e., unseen cases). In fact the initial 212 cases should not really be considered as "training data" until they have been processed by the system and perhaps a rule added. For example in Fig. 8(c), in processing the first 100 cases, 22 errors occurred and a rule was added for each error as it occurred. The remaining 112 cases had not yet been used for training; however, in

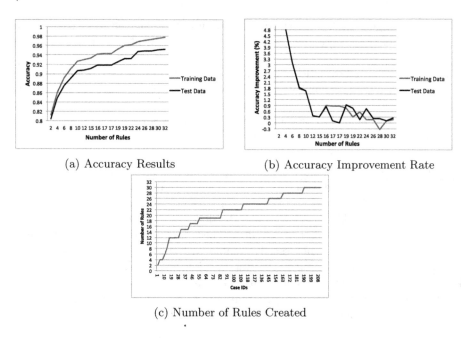

(a) Accuracy Results   (b) Accuracy Improvement Rate

(c) Number of Rules Created

Fig. 8. Experiment results

the following discussion of accuracy, for simplicity, all 212 cases eventually used for training are used in assessing accuracy on training data.

Observations:

- Figure 8(a) shows that, during the training phase, the performance of the system improves as more sentences are processed and more rules are added. The performance improves rapidly at the early stage of the training and gradually plateaus. At the end of the training cases, 32 rules had been added and the accuracy was 98%. The accuracy is not 1.0 because when a new rule is added only the case for which the parent rule is added is checked; however other cases in the training data processed by this rule, might now be incorrect. The results demonstrate that even checking one case per rule provides very robust validation.

  The performance on the test data similarly improves rapidly as more rules are added to deal with training data errors. The accuracy on the test data and training data is very similar when low numbers of rules have been added, because in fact most of the "training data" is as yet unseen; however, as more and more of the training data is actually used for training, the performance on the test data is only slightly less than the training data, 96% vs 98%.

- Figure 8(b) shows that a new rule had bigger impact on the performance of the system at the early stage of the experiment. Then, as more and more rules were added their impact tailed off. This is because common and repeated errors are fixed earlier on, leading to substantial improvement in terms of the accuracy measure. The overall trend of the graphs shows that the improvement brought by each rule converges at a low volume, because the errors left to be fixed at the late stage of the experiment are most likely only applicable to unique and less common cases.

- Figure 8(c) shows how quickly rules are added as the cases are processed. Rules are initially added frequently as exceptions are discovered. Since the exceptions are subsequently re-used to handle new sentences, less new rules are required. Eventually, as the other graphs show, sufficient rules have been added to the system to handle most new sentences.

On Average Time Per Rule Creation: Because creating rules requires understanding of the NLP parse tree structure, we also have looked at the average time taken per rule creation. This was done by asking three users (one expert, two non-experts) to use the system. The expert who was the developer of the system had around 12 hours of experience using the rule editor. The non-experts were aware of the purpose of the system, but not necessarily familiar with the parse tree structure. They had around 2 hours of training on the system. They did not participate in building knowledge bases, but experimented with the system adding around 5-10 rules. The expert reported that it took less than 2 minutes to look at a case and create a rule if needed. For the non-experts, it took around 2 minutes to create a rule.

7   Conclusion

In this paper, we presented an RDR-based knowledge acquisition system to extract action/data pairs from text documents (specifically recipes) that contain process instructions (i.e., step-by-step guides). Although omitted for clarity, our system also provides a rule-based component that improves the POS tagging itself, which in turns improves the action-data extraction process overall. The performance of the system over a test data set showed that even with relatively short training, we could obtain 96 % accuracy on unseen data.

Our immediate future work includes performing more experiments with multiple users, using the same dataset and ground truth. This is not only to show that the results can be repeated by different users, but also to gain deeper insights into the relationships between the number and quality of the rules and the performance improvement of the system. In machine learning research, repeat experiments with different randomisations of training and test data, are used to avoid spurious results due to differences between training and test data. The results here are clearly not due to significant differences between training and test data as the increase in accuracy as cases are processed and rules are added is very similar for both training data (including cases from the 212 not yet processed in training) and test data (cases 213 to 317).

Another important aspect of the work is to design an approach that could assist users with writing the rules over the parse trees, as this requires a high level of understanding of the phrase representations. The email classification system built by [4,13] shows that a purpose-designed rule editor interface that facilitates a simple point-and-click style can help the rule building process. The extensive evaluation plan we outlined above could help us identify critical and common patterns in the generated rules. Such patterns could be the basis for the design of a more intuitive rule editor.

References

1. Compton, P., Jansen, R.F.: A philosophical basis for knowledge acquisition. Knowl. Acquisition **2**(3), 241–258 (1990)
2. Dani, M., Faruquie, T., Garg, R., Kothari, G., et al.: A knowledge acquisition method for improving data quality in services engagements. In: Services Computing, pp. 346–353. IEEE (2010)
3. Hajimirsadeghi, S.A., Paik, H.-Y., Shepherd, J.: Social-network-based personal processes. In: Bae, J., Suriadi, S., Wen, L. (eds.) AP-BPM 2015. LNBIP, vol. 219, pp. 155–169. Springer, Heidelberg (2015)
4. Ho, V., Wobcke, W., Compton, P.: EMMA: an e-mail management assistant. In: Intelligent Agent Technology, pp. 67–74. IEEE (2003)
5. Hsu, J.O., Paik, H., Zhan, L.: Similarity search over personal process description graph. In: Wang, J., Cellary, W., Wang, D., et al. (eds.) WISE 2015, Part I. LNCS, vol. 9418, pp. 522–538. Springer, Heidelberg (2015). doi:10.1007/978-3-319-26190-4_35
6. Kim, M.H., Compton, P.: Improving the performance of a named entity recognition system with knowledge acquisition. In: ten Teije, A., Völker, J., Handschuh, S., Stuckenschmidt, H., d'Acquin, M., Nikolov, A., Aussenac-Gilles, N., Hernandez, N. (eds.) EKAW 2012. LNCS, vol. 7603, pp. 97–113. Springer, Heidelberg (2012)
7. Kim, M.H., Compton, P., Kim, Y.S.: RDR-based open IE for the web document. In: K-CAP, pp. 105–112. ACM (2011)
8. Kim, Y.S., Park, S.S., Deards, E., Kang, B.H.: Adaptive web document classification with MCRDR. In: ITCC, vol. 1, pp. 476–480 (2004)
9. Kwok, R.B.H.: Translations of ripple down rules into logic formalisms. In: Dieng, R., Corby, O. (eds.) EKAW 2000. LNCS (LNAI), vol. 1937, pp. 366–379. Springer, Heidelberg (2000)
10. Manning, C.D., Surdeanu, M., Bauer, J., Finkel, J., Bethard, S.J., McClosky, D.: The stanford CoreNLP natural language processing toolkit. In: ACL (Systems Demonstration), pp. 55–60 (2014)
11. Nguyen, D.Q., Pham, S.B., Pham, D.D.: Ripple down rules for part-of-speech tagging. In: Gelbukh, A.F. (ed.) CICLing 2011, Part I. LNCS, vol. 6608, pp. 190–201. Springer, Heidelberg (2011)
12. Richards, D.: Two decades of ripple down rules research. Knowl. Eng. Rev. **24**(2), 159–184 (2009)
13. Wobcke, W., Krzywicki, A., Chan, Y.-W.: A large-scale evaluation of an e-mail management assistant. In: Web Intelligence and Intelligent Agent Technology, pp. 438–442. IEEE (2008)
14. Xu, J., Paik, H., Ngu, A.H.H., Zhan, L.: Personal process description graph for describing and querying personal processes. In: Sharaf, M.A., Cheema, M.A., Qi, J. (eds.) ADC 2015. LNCS, vol. 9093, pp. 91–103. Springer, Heidelberg (2015)

Specialized Review Selection Using Topic Models

Anh Duc Nguyen[✉], Nan Tian[✉], Yue Xu[✉], and Yuefeng Li[✉]

Faculty of Science and Engineering, Queensland University of Technology,
Brisbane, Australia
{a49.nguyen, nan.tian}@hdr.qut.edu.au,
{yue.xu, y2.li}@qut.edu.au

Abstract. Online reviews and comments about a product or service are an invaluable source of information for users to assist them in making purchase decisions. In recent years, the research in review selection has attracted considerable attention. Many of the existing works attempted to identify a number of statistical features related to review text such as word count (Mudambi and Schuff 2010) and hidden relations between these features and review quality by using supervised learning methods such as classification techniques. However, one significant drawback of these works is that they do not take the review content into consideration. A recent work has been proposed to find specialized reviews that focus on a specific feature based on similar words to the feature (Long et al. 2014). In this paper, we propose a topic model based method which selects reviews by considering both similar words and related words from a topic model such as LDA model. The conducted experiment has proven that those related words generated from LDA have a great contribution to the task of finding helpful reviews on a specified feature.

Keywords: Review selection · Topic modelling · LDA · Related words

1 Introduction

Online reviews provide valuable information for customers to assist them in making purchase decision. Customers prefer to read comments to get a full overview of their target product before deciding whether they will buy it (Dellarocas et al. 2007). However, the explosive proliferation of reviews in the Internet is also a headache for users. For example, Yelp[1], a review website where customers can leave their reviews for local businesses, has monthly average of 89 million visitors and more than 90 million reviews were written by Yelpers by the end of 2015. It is extremely difficult for ordinary users to sift through such overwhelming reviews to get a general view of their interested product or business. Thus, many websites allow users to vote reviews based on their personal experience in the form of "150 out of 250 people found this review helpfulness", or provide overall rating for the business from 1 to 5 stars. Such voting information can be utilized to determine the review helpfulness in order to save users'

---

[1] http://yelp.com.

© Springer International Publishing Switzerland 2016
H. Ohwada and K. Yoshida (Eds.): PKAW 2016, LNAI 9806, pp. 102–114, 2016.
DOI: 10.1007/978-3-319-42706-5_8

time and effort in finding useful reviews. Although review helpfulness undoubtedly provides general information about the quality of reviews for users, it fails to provide information about some particular features of the business. Feature is an attribute of the product or business that has been mentioned in reviews. For example, "food" and "price" are features found in the sentence "the food of this restaurant is quite spicy but the price is an advantage". It is easy to see that reviews having high ranking of helpfulness may not cover features that the readers really interested in. Several e-commerce websites, such as Trip Advisor[2] have already integrated the feature ratings for each product but a large number of customers do not provide feature ratings in their reviews. In order to address this problem, some works tied to predict feature rating, through sentiment-oriented classification. The work proposed by Hu and Liu (2004) is a pioneer in statistical summary of opinion (negative and positive) on each feature in reviews. However, such works only focus on classifying the semantic opinion for each feature, but ignoring the summarization of the whole reviews. In fact, online users may still prefer to read the whole content of reviews to have a vivid picture of the product. Driven by this motivation, the recent approach by Long et al. (2014) proposed to select specialized reviews in which a single feature has been comprehensively discussed. However, the method only examines those similar words of the target feature without considering the related words which have a close relationship to the target feature in the review context. Take the business St Francis American restaurant in Yelp website for example, the relationship between word "bread" and "goat" to feature "cheese" cannot be found by Long's method. In fact, "goat cheese with bread" frequently appear together in review collection and is in fact one of popular dish of the restaurant. If we consider "cheese" is a feature of the restaurant, Long's method cannot find "goat" and "bread" as related words of feature "cheese".

On the other hand, topic modeling Latent Dirichlet Allocation (LDA) has attracted significant attention in recent works (Blei and McAuliffe 2007; Blei et al. 2003; Gao et al. 2013; Griffiths et al. 2005). LDA model can learn and discover topics which are a cluster of words that tend to co-occur in a subset of documents. In this paper, we propose to employ topic modeling LDA to find the set of all related words to a target feature. Those related words, together with similar meaning words and dependent words of the target feature, assist in finding a number of helpful and comprehensive reviews based on the information distance calculation proposed by Long et al. (2014). One interesting thing to notice is that the related words of the target feature found by LDA may also be other features of the business. For example, "bread" is a related word of the feature "cheese" but itself also an individual feature of the business in the above example. Our proposed method aims to accurately identify the relevant information about the target feature in order to select those reviews that select reviews by replying those related words so the returned reviews not only discuss about the target feature but also discuss other related features. Our approach, therefore, contributes to review selection field in term of review comprehension and helpfulness.

---

[2] http://www.tripadvisor.com.

2   Related Work

Our work is related to the existing work that addresses the problem of finding a number of reviews for a given feature based on the criteria of the helpfulness and comprehension. In the following of this section, existing works on review selection will be presented.

2.1   Review Selection Based on Review Quality

Some researchers proposed to assess the helpfulness or quality of the reviews (Kim et al. 2006; Liu et al. 2007). Kim et al. (2006) attempted to utilize SVM regression models to automatically predict the review helpfulness. In order to train the model, a number of features have been defined such as structural, lexical, syntactic, and sentiment aspects of the reviews.

Liu et al. (2007) observed three types of biases in the work of Kim et al. (2006), which includes imbalance vote bias, winner circle bias and early bird bias. They proposed a supervised method to automatically detect low-quality reviews based on the annotated ground-truth.

Hong et al. (2012) attempted to improve the earlier success of Kim et al. (2006) in accuracy rankings of system by learning user preferences (information needs fulfilled by product reviews, the credibility of reviews, and each reviews' consistency with the mainstream opinion of the product) within the regression model. The results show that the automatic helpfulness voting performance is improved by taking user preferences into consideration.

2.2   Review Selection Based on Features

Selecting reviews based upon their predicted helpfulness or quality has a number of drawbacks. First, the top ranked reviews may contain redundant information, which has a low coverage of product features. For instance, users may find that most of the top selected reviews of a restaurant only talk about feature "food" but nothing about other features such as "atmosphere", "price", etc. Second, existing review quality prediction methods usually require considerable time and resources in labeling data and training the system. In terms of the shortcomings of the traditional supervised approaches, (Kim et al. 2006; Hong et al. 2012) and Tsaparas et al. (2011) proposed to select a small number of reviews to represent the whole review corpus. Their method is to assure that the generated review set covers all product features and corresponding viewpoints (positive, negative or neutral). They simply formulated this selection as a maximum coverage problem.

The method of Tsaparas et al. (2011) fails to reflect the proportion of positive and negative opinions on each feature. Thus, Lappas et al. (2012) proposed a method to take such review corpus characteristic into consideration. More specifically, their work is to generate a subset of reviews that accurately reflects opinion distribution for each feature in the underlying review corpus. The selected reviews can help users to know the strength and weakness of the product by reading a compact body of reviews. The problem of Lappas et al. (2012) and Tsaparas et al. (2011) works is that they only focus on the overall utility of the selected reviews, while the quality of each individual review has been ignored.

Recently, Zhang and Liu (2011) and Long et al. (2014) have proposed a novel method in this line of work. The authors argue that most review selection approaches do not consider personal interest. In other words, some users may be interested in some particular features rather than all features. Therefore, those reviews that intensively discuss the feature they are interested will be more useful for them. (Long et al. 2014). Given a specified feature and a review collection, their model extracts a set of similar words to that feature based on Google code of length (Cilibrasi and Vitanyi 2007). Then the authors use Kolmogorov complexity and information distance to calculate the amount of information based on these related word set. The most specialized review on a feature is the one with minimal information distance. However, one significant drawback of this method is that similar words of those core feature words are not necessarily related to certain contexts that the core features are discussed. Take feature "star" for example, when using Google distance to find similar words of feature "star", words "genius", "lead", "stellar" returned. The word "star", in the context of restaurant, indicates the ranking of restaurant and clearly does not have the similar meaning to word "genius", "lead" or "stellar'. Another shortcoming of this method is that the selected specialized reviews may be not helpful to users. As discussed above, reviews that cover more features tend to be more helpful reviews (Tsaparas et al. 2011). If we only focus on finding reviews that discuss one particular feature, the helpfulness can be negatively affected. It is also the fact that professional users tend to write down their opinion on a group of related features of the product or business. Therefore, if we find specialized reviews that discuss only one feature, we may miss some other high-quality reviews.

In order to tackle this problem, we propose a topic modeling based method to find a group of related features of the target feature for assisting in review selection.

3 The Proposed Approach

Our review selection approach aims to selects reviews that discuss not only a targeted feature but also related features. According to Long et al. (2014), the amount of information discussed on the targeted feature in a review determines the comprehension of that review on the targeted feature. They claim that the amount of information can be calculated by words having similar meaning to the feature. As mentioned in the Introduction section, Long's method does not discover other related words or features closely related to the target feature. We, therefore, propose to use topic modelling technique to find those related words. In this paper, we present an approach which takes a review collection and a target feature as the input and the output is a set of highly-ranked reviews. Figure 3 provides an overview of our proposed approach. We introduce the method to extract similar words and dependent words in Sect. 3.1. Section 3.2 discusses the extraction of related words by using topic modelling. Our review selection method is discussed in Sect. 3.3.

3.1   Identifying Similar Words

Apart from the feature words, other synonym or similar words can be used by users to refer to the same feature. For instance, the words "atmosphere" and "ambiance" are interchangeable when users discuss restaurant's atmosphere. Thus, word "ambiance" should be included in the list of similar words of feature "atmosphere".

In addition to the similar words, sentiment words in sentences are used to describe the features. For example, in the three sentences shown in Fig. 1, sentiments such as "amazing", "cosy", "chill", "modern" and "welcoming" are used to describe the feature "atmosphere" and its synonyms "air" and "ambience". These words contribute to the user's opinion about "atmosphere" and should be considered to be relevant to the feature "atmosphere". According to Marneffee et al. (2006), these words are called dependent words because they have grammatical dependent relationship with the feature.

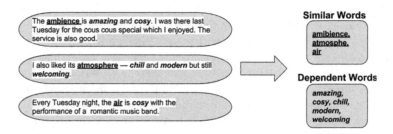

Fig. 1. Similar and dependent words of Feature "atmosphere"

In our approach, we identify features' similar words by using the lexical ontology WordNet (Manna and Mendis 2010). As for the dependent words that are usually adjectives, we employed the distance-based sentiment word extraction method (Lau et al. 2009).

3.2   Extracting Related Words Based on Topic Models

Topic modelling is considered to be the state-of-the-art text mining technique which provides a tool to discover a topic model in large archives of text. The basic idea of topic modelling is that every document is generated by a mixture of topics and a topic is a multinomial distribution over words. Latent Dirichlet Allocation (LDA) proposed by Blei et al. (2003) is currently the most popular approach in generating topic models. Given a collection of documents, LDA can learn and discover topics each of which is represented by a group of words that tend to co-occur in the documents. The probability distribution over words in each topic indicates the importance of each word to the topic and the probability distribution over topics in each document indicates the importance of each topic to the document.

Let $D = \{d_1, d_2, \ldots, d_M\}$ be a collection of M documents. The topic model generated by using LDA consists of topic representations at collection level and topic distributions at document level. At collection level, each topic Z_j is represented by a probability distribution over words, $\phi_j = (\varphi_{j,1}, \varphi_{j,2}, \cdots, \varphi_{j,n})$, $\sum_{k=1}^{n} \varphi_{j,k} = 1$, $\varphi_{j,k}$ is the weight for the kth word. At document level, each document is represented by probability distribution over topics, $\theta_{d_i} = (\vartheta_{d_i,1}, \vartheta_{d_i,2}, \ldots, \vartheta_{d_i,V})$ where V is the number of topics, $\theta_{d_i,j}$ is the probability of topic j for document d_i.

LDA has been widely used in many application domains. In this paper, we propose to apply LDA into review collection to discover feature relationships based on the discovered topics.

3.3   Extracting Topic Related Words

As discussed in Introduction section, Long's method cannot discover related words or features that frequently appear with the target feature because Long's method only uses Google distance or WordNet to find similar words of the target feature. LDA is a probabilistic model that can group related words together into topics. We can use it to find more related words of the target feature. More specifically, we first apply LDA to our review corpus to generate a set of topics. Let each topic be denoted by $Z_i = \{w_1, w_2, ..., w_n\}$, where w_k is k^{th} word assigned to topic Z_i. It is noticed that words in topic Z_i have different weights indicating their different degree of importance for the topic. We then do filtering to remove low-weighted words in the topic by using a minimum threshold of σ. Figure 1 lists remaining words of topic 5 after removing low-weight words. Let Z_i' be i^{th} topic after doing filtering, Z_i' is defined as below.

$$Z_j' = \{w_k | w_{k \in Z_i}, \varphi_{i,k} \geq \sigma\}, \text{ where } \varphi_{i,k} \text{ denote the weight of } w_k.$$

The words in Z_i' are considered related with each other because they are selected to represent the same topic. For a given feature f, if f is a topic word for a topic Z_i, the words in Z_i can be considered topic related words to f.

According to Lappas et al. (2012), reviews that cover more features are considered be more helpful. Our method can find a bag of words that contains not only the target feature but also other topic related words. As shown in Fig. 2, these found topic related words contain some nouns which may also be considered as features of the business. As a result, the set of selected reviews will cover more features and their helpfulness can be improved. One thing to notice is that more related features do not mean that our method no longer focuses on the target feature. The target feature is still the main feature of the review selection process.

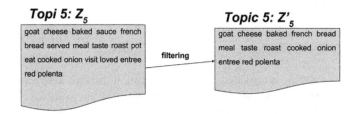

Fig. 2. Topic 5 after removing words having low weight.

3.4   Review Selection

For a given feature f, we want to find a set of reviews R_f, each of which provide the information about the feature. In order to generate the R_f, we need to measure the information which is relevant to the feature f buried in a review. Inspired by the work proposed by Long et al. (2009), we propose to measure the information in a review that relates to a given feature using the Kolmogorov complexity of the feature's similar words and the feature's topic related words.

For an object w, the Kolmogorov complexity of w, denoted as $K(w)$, expresses the information contained in w (Grünwald and Vitányi 2003). Theoretically, the Kolmogorov complexity of w is defined as the length of the shortest effective binary description of A (Grünwald and Vitányi 2003). However, $K(w)$ is not computable in general. Following the idea in (Long et al. 2009), in this paper, we use the relevance of a word w to feature f and Shannon-Fano code (Li and Vitányi 2013) to measure the $K(w)$ relative to f. Given a feature f, the relevance of a word w to f can be measured by the conditional probability $P(w|f) = P(w,f)/P(f)$, where $P(w,f)$ can be approximated by the document co-occurrence of w and f and $P(f)$ can be approximated by document frequency of f, that is:

$$K(w) = -logP(w|f) = -logP(w,f) + logP(f)$$

Let $R = \{r_1, \ldots, r_M\}$ be a collection of reviews, f be a given feature, and $SDW_r(f)$ be a set of similar and dependent words in an individual review r, generated using the method discussed in Sect. 3.1, then a set of similar and dependent words to f in the review collection, denoted as $SDW_R(f)$, can be generated as $SDW_R(f) = \bigcup_{r \in R} SDW_r(f)$. Similarly, let $RW_R(f)$ be a set of related words to feature f in the review collection R, $RW_R(f) = U_{j=1, f \in Z_j}^{V} Z_j'$, generated using the method discussed in Sect. 3.3, then the topic related words to f in an individual review r, denoted as $RW_r(f)$, can be generated as $RW_r(f) = \{w | w \in RW_R(f) \, and \, w \in r\}$.

Accordingly, by combining all the similar, dependent and related words together, we get a set of similar, dependent and related words to feature f in the collection and each individual review. Let $SDRW_R(f)$ and $SDRW_r(f)$ denote the set of similar, dependent and related words to f in the collection and an individual review r, respectively, then $SDRW_R(f) = U_{r \in R} SDRW_r(f)$, and $SDRW_r(f) = SDW_r(f) \cup RW_r(f)$.

The following score is calculated to measure the Kolmogorov complexity of a review r in terms of feature f by calculating the Kolmogorov complexity of the words in other reviews rather than in r:

$$SPE_{r,f} = \sum_{w \in SDRW_R(f) \backslash SDRW_r(f)} K(w) = \sum_{w \in SDRW_R(f) \backslash SDRW_r(f)} logP(f) - logP(w,f)$$

The value of $SPE_{r,f}$ is considered as the information distance between $SDRW_R(f)$ and $SDRW_r(f)$. The less the distance is, the more relevant the words in r to f. Therefore, reviews having the lowest score of $SPE_{r,f}$ are selected as the output of our system.

4   Experiment and Evaluation

4.1   Datasets

In our experiments, we use an industry published review dataset provided by RecSys conference from Yelp in 2013[3]. The datasets include detailed data of over 10,000 businesses, 8,000 check-in sites, 40,000 users, and 200,000 reviews from the Phoenix, AZ metropolitan area. In recent years, Yelp datasets have been widely used in the research of opinion mining and recommendation system.

All reviews can be extracted from the file yelp_training_set_review.json provided by Yelp website (see Footnote 3). The structure of each review is in the form of JSON format and described as below:

```
{
  'type': 'review',
  'business_id': (encrypted business id),
  'user_id': (encrypted user id),
  'stars': (star rating),
  'text': (review text),
  'date': (date),
  'votes': {'useful': (count), 'funny': (count), 'cool': (count)}
}
```

In order to choose suitable datasets with sufficient amount of reviews and sufficient average amount of words in each review, we have done some statistics for the whole datasets and choose those datasets that meet the criteria for as follows:

- Dataset having sufficient number of reviews (at least 200 reviews).
- Reviews having sufficient number of words (average number of words should be greater than 130 words).

By following the above criteria, we randomly select four datasets in different categories of business for our experiments. Table 1 provides the detailed information about each chosen dataset.

Table 1. Dataset Information

Category	Businesses name	Number of reviews	Average number of words in a review
American Food	St Francis Restaurant	461	165
Breakfast & Brunch	Lo-Lo's Chicken & Waffles	357	139
Pizza	Cibo	594	133
Pizza	Papago Brewing	236	142

---

[3] https://www.kaggle.com/c/yelp-recsys-2013.

4.2   Experiment Procedure

The whole procedure of our experiment is depicted in Fig. 3. We used four datasets described in Sect. 4.1 for our experiment. Each of these dataset was the input of our model and the output is the set of high-ranking reviews. The method proposed by Long et al. (2014) was chosen as the baseline for comparison. The returned set of reviews from our method and the baseline model were analyzed, evaluated and compared.

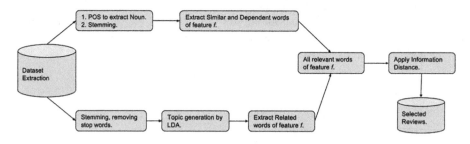

Fig. 3. Experiment procedure

4.3   Result Analysis and Evaluation

4.3.1   Evaluation of Helpfulness

Yelp website allows users voting each review to indicate if it is helpful from their perspective. Each review is associated with votes in three different categories namely "useful", "funny" and "cool". We use the votes in "useful" category to determine the usefulness of review text. The helpfulness score of each review is calculated by the ratio of count of usefulness of the review and the total count of usefulness of all review.

Our method selects reviews based on the input of the single feature. According to Hu and Liu (2004), features are attributes of a product or business that customers have commented on in textual reviews. Nouns which frequently appear in the reviews about a product are more likely to be the features of the product. Therefore, we applied pattern mining Apriori algorithm (Agrawal and Srikant 1994) to find a list of frequent words occurring in the review collection. Minimum support (frequency of words appearing in review sentences) of 1 % is used as the threshold to generate frequent words. However, not all of those frequent words are genuine features. In our experiment, the human annotators examined those frequent words and removed words that are clearly meaningless. After the pruning, the top 10 frequent words are chosen as 10 features for testing. More specifically, for each feature, both testing methods are employed to extract 30 reviews. We calculate the average helpfulness score of selected reviews from both our method and the baseline for performance comparison. The experimental results for four datasets are given in Tables 2, 3, 4 and 5.

Table 2. Total score of selected review of features of St Francis Restaurant

Features	Baseline	Our Approach
Average	0.1581	0.1614

Table 3. Total score of selected review of features of Lo-Lo's Chicken & Waffles

	Baseline	Our Approach
Average	0.1920	0.2040

Table 4. Total score of selected review of features of Cibo

	Baseline	Our Approach
Average	0.1249	0.1304

Table 5. Total score of selected review of features of Papago Brewing

	Baseline	Our Approach
Average	0.1400	0.1620

According to the comparisons, we can see that the helpfulness scores of selected reviews generated by our method are always higher than that of the baseline for most of input features. Therefore, it is evident that our approach can improve the performance of Long's approach.

4.3.2   Feature Comprehension Evaluation

There is no standard way to evaluate the comprehension of discussion of a specific feature in each review. Thus, we ask two annotators to read the 30 selected reviews and decide that how comprehensively the reviews discuss the specified feature. Table 6 lists top 30 returned reviews and Table 7 lists the non-overlapping reviews of the baseline and our method for input feature "food". According to Table 6, there are many overlapping reviews in the top 30 reviews generated from our method and the baseline. Therefore, instead of examining all selected reviews, we only analyze those non-overlapping reviews returned from our system and the baseline. The annotators read each non-overlapping reviews in Table 7 to give the score of comprehension to each review. The score of those reviews in term of comprehensive discussion given by annotators will be used for comprehension evaluation.

Table 6. Top 30 selected reviews returned by Long and our approach for feature "food"

Baseline	Our method
Review 161, 60, 189, 214, 124, 141, 298, 187, 255, 254, 66, 335, 323, 320, 304, 23, 249, 92, 330, 19, 122, 333, 341, 34, 76, 95, 132, 94, 153, 186	Review 189, 60, 214, 323, 254, 161, 124, 330, 187, 23, 226, 122, 304, 320, 141, 153, 129, 335, 66, 296, 186, 132, 19, 34, 333, 249, 199, 274, 298, 251

Table 7. Non-overlapping reviews returned by Long and our approach for feature "food"

Baseline	Our method
Review 255, 92, 341, 76, 95, 94	Review 251, 274, 199, 296, 129, 226

The criteria for comprehensive discussion of a feature depends on the number of sentences discussing that feature and its related features in the review. As shown in Tables 8, 9, 10, and 11, the total score of the selected reviews generate by our method are better than the results of the baseline. Therefore, our method can improve the performance of the baseline in term of comprehensive discussion.

Table 8. Total score of selected review of St Francis Restaurant in term of comprehension

	Baseline	Our Approach
Average	12.23	12.25

Table 9. Total score of selected review of Lo-Lo's Chicken & Waffles in term of comprehension

	Baseline	Our Approach
Average	9.25	12.40

Table 10. Total score of selected review of Cibo in term of comprehension

	Baseline	Our Approach
Average	11.35	14.25

Table 11. Total score of selected review of Papago Brewing in term of comprehension

	Baseline	Our Approach
Average	10.24	10.95

The improvement can be explained by related words of the target feature. As shown in Table 12, for input feature "food", our method generates more related words of the feature "food" such as feature "chicken", "buttery", "waffle", "cheese", "okra". Those related words or features cannot be found by the baseline method. Therefore, the probability of getting comprehensive reviews for feature "food" by using our method is certainly higher than the baseline method.

Table 12. Similar, dependent and related words of feature "food"

Baseline	Our method
food, dish, good, outside, great, awesome, delicious, fast.	food, dish, good, outside, great, awesome, delicious, fast chicken, buttery, waffle, cheese, okra, fish, platter, bean, cornbread.

5   Conclusion

Review selection is one of the hot research topics due to its significance in decision making process. Most existing works analyze textual characteristics and opinion mining to select helpful reviews. In recent years, many research focus on selecting reviews that cover comprehensive information about features of the product or business. In this paper, we introduced a method which utilizes LDA to select reviews that focus on one particular feature. LDA topic model helps to discover relationships between the target features and other related words, which assists in extracting more relevant information about the feature. Some of those related words are also the features of the business which are added together with the target feature to create a bag of words. This bag of words containing related features to the target feature is then used to select the most relevant reviews, thereby improving the performance of our method in term of both review helpfulness and comprehension. The conduct evaluation based on 4 business datasets collected from Yelp indicates that our method achieve better performance compared to the baseline model.

References

Agrawal, R., Srikant, R.: Fast algorithms for mining association rules. In: Proceedings of 20th International Conference Very Large Data Bases, VLDB, vol. 1215, pp. 487–499 (1994)

Blei, D.M., McAuliffe, J.D.: Supervised topic models. In: Advances in Neural Information Processing Systems (NIPS) (2007)

Blei, D.M., Ng, A.Y., Jordan, M.I.: Latent Dirichlet allocation. J. Mach. Learn. Res. **3**, 993–1022 (2003)

Cilibrasi, R.L., Vitanyi, P.: The google similarity distance. IEEE Trans. Knowl. Data Eng. **19**(3), 370–383 (2007)

Dellarocas, C., Zhang, X.M., Awad, N.F.: Exploring the value of online product reviews in forecasting sales: the case of motion pictures. J. Interact. Mark. **21**(4), 23–45 (2007)

Gao, Y., Xu, Y., Li, Y.: Pattern-based topic models for information filtering. In: IEEE 13th International Conference on Data Mining Workshops, pp. 921–928. IEEE (2013)

Griffiths, T.L., Steyvers, M., Blei, D.M., Tenenbaum, J.B.: Integrating topics and syntax. In: Advances in Neural Information Processing Systems, vol. 17, pp. 537–544 (2005) (Retrieved)

Hong, Y., Lu, J., Yao, J., Zhu, Q., Zhou, G.: What reviews are satisfactory: novel features for automatic helpfulness voting. In: SIGIR Conference on Research and Development in Information Retrieval, pp. 495 – 504. ACM (2012)

Hu, M., Liu, B.: Mining and summarizing customer reviews. In: Proceedings of the Tenth ACM SIGKDD International Conference on Knowledge Discovery and Data Mining, pp. 168–177. ACM (2004)

Kim, S.-M., Pantel, P., Chklovsk, T., Pennacchiotti, M.: Automatically assessing review helpfulness, pp. 423–430. Association for Computational Linguistics (2006)

Lappas, T., Crovella, M., Terzi, E.: Selecting a characteristic set of reviews. In: 18th SIGKDD International Conference on Knowledge Discovery and Data Mining. ACM (2012)

Lau, R.Y., Lai, C.C., Ma, J., Li, Y.: Automatic domain ontology extraction for context-sensitive opinion mining. In: ICIS 2009 Proceedings, pp. 35–53 (2009)

Li, M., Vitányi, P.: An Introduction to Kolmogorov Complexity and its Applications. Springer Science & Business Media, New York (2013)

Liu, J., Yunbo, C., Chin-Yew, L., Yalou, H., Ming, Z.: Low-quality product review detection in opinion summarisation, pp. 334–342. Association for Computational Linguistics (2007)

Long, C., Zhang, J., Huang, M., Zhu, X., Li, M., Ma, B.: Estimating feature ratings through an effective review selection approach. Knowl. Inf. Syst. **38**(2), 419–446 (2014)

Manna, S., Mendis, B.S.U.: Fuzzy word similarity: a semantic approach using WordNet. In: 2010 IEEE International Conference on Fuzzy Systems (FUZZ), pp. 1–8. IEEE (2010)

Mudambi, S.M., Schuff, D.: What makes a helpful review? A study of customer reviews on Amazon. com. MIS Q. **34**(1), 185–200 (2010)

Grünwald, P.D., Vitányi, P.M.: Kolmogorov complexity and information theory. With an interpretation in terms of questions and answers. J. Logic, Lang. Inf. **12**(4), 497–529 (2003)

Tsaparas, P., Ntoulas, A., Terzi, E.: Selecting a comprehensive set of reviews. In: 17th SIGKDD International Conference on Knowledge Discovery and Data Mining, pp. 168–176. ACM (2011)

Zhang, L., Liu, B.: Identifying noun product features that imply opinions. In: Proceedings of the 49th Annual Meeting of the Association for Computational Linguistics: Human Language Technologies: short papers, vol. 2, pp. 575–580. Association for Computational Linguistics (2011)

Knowledge Acquisition from Network and Big Data

Competition Detection from Online News

Zhong-Yong Chen[(✉)] and Chien Chin Chen

Department of Information Management, National Taiwan University,
No. 1, Section 4, Roosevelt Road, Taipei City 10617, Taiwan, ROC
{d98725003,patonchen}@ntu.edu.tw

Abstract. In this paper, we define a novel problem named competed intention identification of online news. We propose new features to represent the competed intention of the documents. The support vector machine (SVM) is employed to adopt our features to identify the competed intention in the news article. Experimental results demonstrate that the features we designed are effective for identifying the documents with competed intention.

Keywords: Text mining · Document classification · Competition identification · Feature extraction

1 Introduction

The developments of firms depend on the information which decision-makers have. The information includes news and professional analysts' reports [14]. Since analyzing the news on the Internet to facilitate decision-makers in making decisions has been popular for decades [3, 4, 7, 8, 10–14, 17, 18, 20, 23, 24] which ranging from analyzing the implied volatility of firms [10], predicting stock prizes [12, 23], prize trends [11], the behavior of the financial market, the market activity [24], market liquidity [13], and analyzing the relations of companies [12, 17, 18], the news has played an important role in decision-making in recent years [12–14, 17, 18]. Nowadays, enormous news about companies is available on the Internet. Investigators often identify the relations between companies from the news for assisting decision-makers in making decisions [17]. Usually, the relations include cooperation, competition, and co-competition [3, 4]. Especially, the news involving competing viewpoints has been gotten highly attention among the relations [18]. However, with the prevalence of medium digitalization, astronomical news is available on the Internet [14]. As a result, the decision-makers have great difficulties in discovering the competitions of companies from the news and need to spend a great deal of time to digest the news. Hence, in this paper, we propose a set of features to discover the news involving competing viewpoints, and develop a new ensemble method to improve the performance of identifying competed documents. Identifying them can facilitate the decision-makers to filter out the less important news and help them digest the news quickly. In addition, the identification of the news involving competing viewpoints can also help readers find the news of interest [7, 8].

H. Ohwada and K. Yoshida (Eds.): PKAW 2016, LNAI 9806, pp. 117–128, 2016.
DOI: 10.1007/978-3-319-42706-5_9

The paper is organized as follows. Section 2 provides a literature review, and Sect. 3 detail the features we proposed. Experimental results are shown in Sect. 4, and we make concluding remarks in Sect. 5.

2   Related Work

To the best of our knowledge, competed intention identification of the documents is a novel research field. Paek et al. [21] investigated the problem of identifying the importance of newsfeed posts and friends from social medium. They claimed that this is the first research to apply machine learning to build predictive models of newsfeed importance. In their research, social media properties, message text, and shared background information are extracted as features for SVM classifier. The predicted results show that the features are effective for predicting the importance of the news-feed posts and friends. Chen et al. [9] proposed two methods to cope with the iden-tification problem of intention posts in transfer learning. They also claimed that their research is the frontier on intention classification. The first method, namely Feature Selection Expectation Maximization method (FS-EM) which uses Information Gain to select feature from source dataset within each iteration in EM, and examine the features on the target dataset. The second method, namely Co-Classification, is to aggregate two Bayes classifiers learning from different datasets to classify intention posts. Experi-mental results show that the Co-Classification is effective in identifying intention posts across different datasets. However, our research is much different from the above in many respects. First, we focus on the news article instead of the forum post or the post on the social network, the news article length is much longer than the posts. The news articles also contain more noisy and informative information which need to filter out and extract respectively. That makes our problem more difficult. Second, Chen et al. [9] discussed a very specific intention, namely buy intention, in their paper; but we focus on the competed intentions which contain a variety of domains. In difficult domain, the manifestation of competed intention is very different, e.g., in the sports tournament, players can say that *we will beat the other team* or *we are well-prepared for the battle*. In the business events, businessmen can state that *we will take more market share than the other company*. The above examples show that the competed intention exists in different domain and can be displayed in different usage of words.

3   Methodology

In this section, we elaborate on our system architecture. Firstly, we extracted features from the input documents in terms of three different granularity: term level, sentence level, and social network level, respectively. Subsequently, we employed the support vector machine [5] to train the prediction model for each feature set from different level, and then again leveraged the SVM to train a new model by using the predicted values from each classifier to identify the competition intention of the online news, as show in Fig. 1.

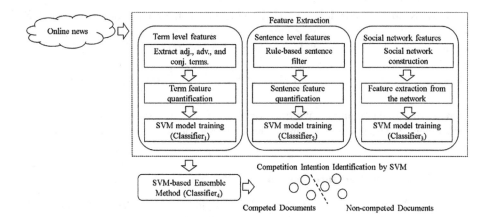

Fig. 1. System architecture

In the next subsections, we detail each component of the system. First of all, we describe the features of different granularity.

3.1   Term Level Feature Selection

Competition is a kind of interactions [6], and using the adjective and adverb terms in the document could be good indicators to identify the interaction of the document [22]. Hence, we extracted the adjective and adverb terms to distinguish the competed documents. Measuring the competition level of the terms is a crucial task for identifying the competed document. We compiled a competing word list, and calculated the similarity between the terms and the list in the WordNet[1] in terms of the approach proposed by Wu and Palmer [25].

Finally, we summed up the similarities between the terms and the list, and calculated the average of the similarities as the document features for the adjective and the adverb terms, as shown in Eqs. (1) and (2), respectively.

$$feature_1 = \sum_{word_i \in d_{adj}} \sum_{word_j \in CL} similarityInWordNet(word_i, word_j) \Big/ |d_{adj}| \times |CL|, \quad (1)$$

$$feature_2 = \sum_{word_i \in d_{adv}} \sum_{word_j \in CL} similarityInWordNet(word_i, word_j) \Big/ |d_{adv}| \times |CL|, \quad (2)$$

where d is a document, d_{adj} (d_{adv}) is a set of adjective (adverb) terms in document d, and CL is the competing word list. The similarityInWordNet($word_i, word_j$) returns the similarity of the term $word_i$ and $word_j$ in the WordNet [25]. $|d_{adj}|$ ($|d_{adv}|$) indicates the number of all the adjective (adverb) terms in the document d. The higher the equation value is, the higher the competed level is.

---

[1] https://wordnet.princeton.edu/.

Hatzivassiloglou and McKeown [16] validated that the conjunction words can be helpful to identify the polarities of terms. As a result, we extracted the conjunction words from the document as its feature, as shown in Eq. (3).

$$feature_3 = \sum\nolimits_{word_i \in d_{conj}} \sum\nolimits_{word_j \in CL} similarityInWordNet(word_i, word_j) \Big/ |d_{conj}| \times |CL|, \quad (3)$$

where $word_i$ is the conjunction word in document d. d_{conj} is a set of conjunction words in document d. The remaining symbols have similar meaning as mentioned above.

3.2   Sentence Level Feature Selection

Zhai et al. [26] dealt with the problem of identifying the evaluative sentences from online postings by designing dependency rules to capture the properties of the postings. Abu-Jbaba et al. [1] also employed dependency rules to identify ideological subgroups in the online forum with specific purpose. Both of the above researches validated that the dependency rules can identify the interactions from the documents correctly. Hence, we extracted the following dependency rules to construct our sentence-level features.

Rule 1: NE₁ (nsubj/nsubjpass/dobj) - verb - NE₂ (nsubj/nsubjpass/dobj). NE_1 and NE_2 mean that the first entity and the second entity in the sentence, and the verb is between them. The *Rule 1* means that the entity (subject/object) directly connects to each other by the verb is a strong relation [26]. If the verb is akin to the competition, it indicates that the two entities have the strong competed relationship. For example, the competed sentence *"The company, i4i, sued Microsoft in 2007, saying it owned the technology behind a text manipulation tool used in Microsoft's Word application."* The rule captured that: *Microsoft (dobj) - sued (verb) - company (nsubj).*

Rule 2: NE₁ (nsubj/nsubjpass) - verb - verb (ccomp/xcomp) - pronoun (nsubj/ nsubjpass). This rule captures the sentences that people express their opinion. As mentioned earlier, the competition is one of the interactions. Expressing the opinion is also a part of the interactions. For example, the sentence *"They can hire as many as they want, but if they don't have customers to come in and shop it doesn't matter, said longtime customer Christa Lamb."* The extracted text units are that: *They (nsubj) - hire (verb) - said (ccomp) - Christa Lamb (nsubj).*

Rule 3: NE₁ (nsubj/nsubjpass)-verb. This rule extracts all the sentences with the verb. It also magnifies the effect of the *Rule 1* and *Rule 2*.

Rule 4: NE₁ (nsubj/nsubjpass)-verb - verb (ccomp/xcomp) – NE₂ (nsubj/nsubjpass). This rule is similar to the Rule 2, but it considers two named entities.

Rule 5: NE₁ (nsubj/nsubjpass) -verb - verb (conj_but) – NE₂ (nsubj/nsubjpass). This rule concerns that the two entities holding different opinion. For example, the sentence *"The NCAA had argued that the 'procompetitivebenefits' of prohibiting athletes from sharing in the multibillion-dollar collegiate sports industry justified the long-held policies, but Wilken disagreed.*" The extracted results are that: *NCAA (nsubj) - argued (verb) - disagreed (conj_but) – Wilken (nsubj).*

Rule 6: NE₁ (nsubj/nsubjpass)-verb – NE₂ (iobj/prep_against/prep_in). The *Rule 6* is similar to the *Rule 1*, but the relationship of the *NE₂* is very different. The main goal of this rule is to capture the intention behavior. For example, the sentence *"The two spurned bidders, the Cultrale Group and the Safra Group, both of Brazil, wrote in a securities filing that Chiquita shareholders should vote against the Fyffes deal and in favor of adjourning the special meeting."* The matched text units are that: Chiquita shareholders *(nsubj) - vote (verb) – Fyffes deal (prep_against).*

Each document can construct the representative vector in terms of the six rules. When we construct the vectors for every document, we need to aggregate them as a feature for the document. We divided the training documents into two parts: the document with competition intention and without competition intention. Then, we construct the vector for each document, whose entry is the number of matching the rules, in the training data, and calculated the centroid for the documents with competition intention and without competition intention, respectively. For a given document (testing document), it will calculate the cosine similarities [19] between the centroids for each part, and then the feature value will be aggregated by the following Eq. (4).

$$feature_4 = \text{cosineSimilarity}(d_i, centroid_p) - \text{cosineSimilarity}(d_i, centroid_{np}), \quad (4)$$

where $centroid_p$ ($centroid_{np}$) indicates the centroid calculated by averaging each vector constructed from the rules with competition intention (without competition intention). The cosineSimilarity function returns the cosine similarity value between the testing document d_i and the $centroid_p$ ($centroid_{np}$). If the value of Eq. (4) is positive, the document d_i has the competition intention tendency, and vice versa. We employed the Stanford dependencies tool[2] to identify the relation of the targets (Entities) mentioned above.

As mentioned earlier, we compiled a competing word list. The extracted sentences by the above rules will be calculated the association between the verb in the sentence and the list we compiled through the WordNet in terms of the approach proposed by Wu and Palmer [25], and then averaging all of the pairs (the verbs in every extracted sentence and the words in the list) as the feature value, as shown in Eq. (5).

$$feature_5 = \sum_{word_i \in ES_v} \sum_{word_j \in CL} \text{similarityInWordNet}(word_i, word_j) \Big/ |ES_v| \times |CL|, \quad (5)$$

where ES_v is a set of the verbs extracted from the sentences in a document.

In addition, we also take the number of matching the rules as a feature. Intuitively, if the document contains a large number of the sentences matching the rules, the document has the propensity to the competition intention. This feature is calculated by Eq. (6).

---

[2] http://nlp.stanford.edu/software/stanford-dependencies.shtml.

$$feature_6 = |ES_d|/|S_d|, \tag{6}$$

where ES_d is a set the sentence matching the rules in the document d, and S_d is a set of the sentences in the document d.

3.3   Social Network Feature Selection

Constructing the network from the news has been getting attention in recent years, e.g., prediction for the company revenue relation [17], and the competitor relationship [18]. Here, we construct the social network G, the topic network, from a set of documents about a topic. Let $G = \{P, E\}$, where $P = \{p_1, p_2, \dots, p_M\}$ is a set of person names which were extracted from the documents, and $E = \{(p_i, p_j)\}$ is a set of edges, whose connections are established by the co-occurrence of the person name p_i and p_j in the same sentence.

After constructing the social network, we calculated the betweenness value of each node which is the shortest path between all the pairs of the nodes in the network through this node. Given a document d, we extracted the betweenness value of the nodes (person names) mentioned in this document, and then averaged them as the feature value of the document as follows.

$$feature_7 = \sum\nolimits_{p_i \in P_d} \text{betweenness}(p_i) \Big/ |P_d|, \tag{7}$$

where P_d is a set of nodes in the document d, and $|P_d|$ is the number of nodes in the document d. The betweenness(p_i) returns the betweenness value of the node p_i in the network G. The smaller average betweenness value indicates that there may exists a coherent community in the document.

In addition, Jaccord coefficient [15] can also represent the dense of the community, which calculates the ratio of the neighbor overlap between two nodes in the network G, as shown in Eq. (8).

$$feature_8 = \sum\nolimits_{p_i, p_j \in P_d} \text{jaccordCoefficient}(p_i, p_j) \Big/ |P_d| \times |P_d|, \tag{8}$$

The degree of the nodes is popular to represent the importance in the network [2, 17, 18], we also calculated the average value of each node in the document from the topic network in Eq. (9).

$$feature_9 = \sum\nolimits_{p_i \in P_d} \text{degree}(p_i) \Big/ |P_d|, \tag{9}$$

where the function degree(p_i) returns the degree of the node p_i in the topic network, and the other symbols are identical to the above mentioned. We will show the experiments to demonstrate the effectiveness of each feature later.

3.4   SVM-Based Ensemble Method

As shown in Fig. 1, we employed SVM to train three models in terms of term level, sentence level, and social network level, to produce the predicted values as new features for each document, and again use the SVM to re-train and predict the competition intention of the documents. We illustrate an example as follows. Given a document d_1, the feature vectors of d_1 are [$feature_1, feature_2, feature_3$], [$feature_4, feature_5, feature_6$], and [$feature_7, feature_8, feature_9$] extracted from different level, respectively. Hence, we have three models trained by SVM, and they would output three predicted values pre_1, pre_2, and pre_3. After that, we leverage the predicted values to construct a new feature vector [pre_1, pre_2, pre_3] for document d_1, and again the feature vector would be trained by SVM to predict the competed intention of the document d_1, as shown in Fig. 2.

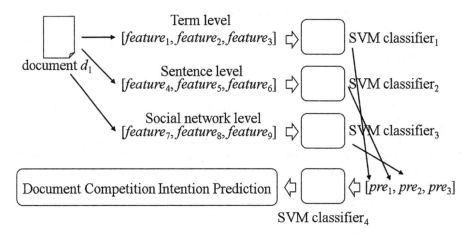

Fig. 2. SVM-based ensemble method

4   Experiments

In this section, we introduced the dataset we compiled, and the effectiveness of the features.

4.1   Dataset

We collected 7 business topics from Google news, each of which comprises of 123, 74, 48, 150, 118, 135, and 92 documents. When we collected the topics, we made sure that there are competition intentions inside the topics. We also asked two linguistic experts to label the topic documents as the ground truth, and the kappa value is good enough to conduct the experiments. The details of the business topics are listed in Table 1.

In Table 1, the column of the Date indicates that the duration of publishing the topic documents. The Topic Description indicates that the topic title which was given by the journalists. To evaluation the experiment result, we employ the metric called

Table 1. The statistics of the dataset

ID	Date	Topic description	# of documents
T_1	2010/7/18 – 2010/7/22	Smartphone manufactures deny Apple's reception claim	123
T_2	2010/8/4 – 2010/8/6	Google-Verizon deny tiered-web deal report	74
T_3	2010/1/13 – 2010/1/15	Google ends four years of censoring the Web for China	48
T_4	2011/5/27 – 2011/6/5	IMF meeting to select a new president	150
T_5	2011/6/6 – 2011/6/10	2011 OPEC meeting to set oil production quotas	118
T_6	2011/6/6 – 2011/6/11	Greek Financial Crisis	135
T_7	2011/6/9 – 2011/6/16	Microsoft and i4i lawsuit over patent violation	92

accuracy [19], which is the ratio of the correctly classified documents to the whole documents. For subsequent experiments, we compiled a competing word list as shown in Table 2. For evaluating the features we extracted, we leveraged the SVM (Support Vector Machine) as our classifier to identify the competition intention of the documents. In addition, we used the libsvm package [5] to implement the SVM. In the subsequent experiments, we used the 7-fold cross-validation to demonstrate the performance of our method.

Table 2. Competing word list

support	share	criticize	campaign
member	approve	rival	abuse
push	benefit	damage	strike
agreement	partner	rape	reject
help	consensus	fight	defend

4.2 Feature Evaluations

As mentioned in Sect. 3, we extracted the features from different level of the document. To demonstrate the effectiveness of the features in each level of the document, we conducted the experiments on term-level, sentence-level, and the social network level, respectively.

In Table 3, we show the experimental results of using term level features to train the model, and leveraged the model to predict the competition intention of the topic documents. Obviously, the $feature_1$, which indicates the similarities between the adjectives of the documents and the competing word list in the WordNet, dominates the influences of the adverbs and the conjunctions. This phenomenon reflects that the topic documents usually convey competition intention with the adjectives, and suggests that the adjectives are good indicators for detecting the competition intention within the documents. For example, in the topic T_1, Steve Jobs always mentions the term "unacceptable" to fight back the accusations of other companies.

Table 3. Term level feature evaluation

	Macro-Accuracy	Micro-Accuracy
$feature_1$	**0.676936**	**0.635958**
$feature_2$	0.660119	0.610698
$feature_3$	0.658095	0.610698
$feature_1 + feature_2$	**0.676936**	**0.635958**
$feature_1 + feature_3$	**0.676936**	**0.635958**
$feature_2 + feature_3$	0.660215	0.612184
$feature_1 + feature_2 + feature_3$	**0.676936**	**0.635958**

In Table 4, we found the experimental result is similar to Table 3 that only one feature dominates the classification performance. Apparently, the salience feature is the $feature_4$ which takes the sentence structure into consideration. As mentioned in Subsect. 3.2, the $feature_4$ is to calculate the difference of cosine similarities between the sentence structure vector of the testing document, and the centroid of the training document with competed intention and without competed intention, respectively. The performance of the $feature_4$ reflects that the documents with competed intention and without competed intention have different sentence structure. In addition, we observed that the competed documents contain more human interaction than the non-competed documents. The finding is coherent with the saying that the competition is a kind of interaction [6].

Table 4. Sentence level feature evaluation

	Macro-Accuracy	Micro-Accuracy
$feature_4$	**0.661407**	**0.612184**
$feature_5$	0.657142	0.609212
$feature_6$	0.659623	**0.612184**
$feature_4 + feature_5$	**0.661407**	**0.612184**
$feature_4 + feature_6$	**0.661407**	**0.612184**
$feature_5 + feature_6$	0.657142	0.609212
$feature_4 + feature_5 + feature_6$	**0.661407**	**0.612184**

In Table 5, we represented the results of employing the social network level features to train the model for predicting the competed documents. The combination of $feature_8 + feature_9$ outperforms other combinations under the social network feature set. The $feature_8$ indicates that the co-occurrence of the node (person) and its neighbors (persons) in the same sentence, and the $feature_9$ indicates that the dense degree of the node's neighbors. In other words, if the larger the value of the $feature_8 + feature_9$ is, the more possibility the coherent group exists in the network. This phenomenon tells us that a dense group of persons in the social network constructing from the topic documents may be a good indicator for identifying the competition intention within the documents.

Table 5. Social network level feature evaluation

	Macro-Accuracy	Micro-Accuracy
$feature_7$	0.657142	0.609212
$feature_8$	0.657142	0.609212
$feature_9$	0.657986	0.609522
$feature_7 + feature_8$	0.675006	0.634473
$feature_7 + feature_9$	0.657142	0.610698
$feature_8 + feature_9$	**0.702932**	**0.677563**
$feature_7 + feature_8 + feature_9$	0.672072	0.656761

Finally, we aggregate the features with the best performance in each feature set to achieve a better prediction performance. We employed the predicted values of the three different level features, each of which has best performance in its level, as the SVM features, to predict the competed documents. Table 6 shows the performance of the SVM-based ensemble method, and this method did improve the accuracy of identifying the competed documents.

Table 6. SVM-based ensemble method evaluation

	Macro-Accuracy	Micro-Accuracy
SVM-based ensemble method	**0.704143**	**0.679049**
$feature_8 + feature_9$	0.702932	0.677563
$feature_1$	0.676936	0.635958
$feature_4$	0.661407	0.612184

5   Conclusions

In this paper, we defined a new problem named competition intention identification of the online documents. Novel features were proposed and extracted from the documents based on different perspective to train the model for generating the predicted values. Furthermore, we proposed a new SVM-based ensemble method to aggregate the predicted values for predicting the competed documents. Experimental results showed that the features and SVM-based ensemble method are effective for identifying the competition intention of the online documents.

Acknowledgement. We are grateful to the anonymous reviewers for insightful comments. This research was supported by the Ministry of Science and Technology of Taiwan under grant MOST 103-2221-E-002-106-MY2.

References

1. Abu-Jbara, A., Diab, M., Dasigi, P., Radev, D.: Subgroup detection in ideological discussions. In: Proceedings of the 50th Annual Meeting of the Association for Computational Linguistics, ACL 2012, pp. 399–409 (2012)
2. Bao, S., Li, R., Yu, Y., Cao, Y.: Competitor mining with the web. IEEE Trans. Knowl. Data Eng. **20**, 1297–1310 (2008)
3. Bengtsson, M., Kock, S.: Coopetition in business networks—to cooperate and compete simultaneously. Ind. Mark. Manage. **29**(5), 411–426 (2000)
4. Bengtsson, M., Kock, S.: Cooperation and competition in relationships between competitors in business networks. J. Bus. Ind. Mark. **14**(3), 178–194 (1999)
5. Chang, C.-C., Lin, C.-J.: LIBSVM: a library for support vector machines. ACM Trans. Intell. Syst. Technol. **2**, 1–27 (2011)
6. Chase, J.M., Abrams, P.A., Grover, J.P., Diehl, S., Chesson, P., Holt, R.D., Richards, S.A., Nisbet, R.M., Case, T.J.: The interaction between predation and competition: a review and synthesis. Ecol. Lett. **5**, 302–315 (2002)
7. Chen, C.C., Wu, C.-Y.: Bipolar person name identification of topic documents using principal component analysis. In: Proceedings of the 23rd International Conference on Computational Linguistics, pp. 170–178 (2010)
8. Chen, C.C., Chen, Z.-Y., Wu, C.-Y.: An unsupervised approach for person name bipolarization using principal component analysis. IEEE Trans. Knowl. Data Eng. **24**, 1963–1976 (2012)
9. Chen, Z., Liu, B., Hsu, M., Castellanos, M., Ghosh, R.: Identifying intention posts in discussion forums. In: Proceedings of the 2013 Conference of the North American Chapter of the Association for Computational Linguistics: Human Language Technologies, pp. 1041–1050 (2013)
10. Donders, M.W.M., Vorst, T.C.F.: The impact of firm specific news on implied volatilities. J. Banking Finan. **20**(9), 1447–1461 (1996)
11. Fung, G.P.C., Yu, J.X., Lam, W.: News sensitive stock trend prediction. In: Chen, M.-S., Yu, P.S., Liu, B. (eds.) PAKDD 2002. LNCS (LNAI), vol. 2336, pp. 481–493. Springer, Heidelberg (2002)
12. Geva, T., Zahavi, J.: Empirical evaluation of an automated intraday stock recommendation system incorporating both market data and textual news. Decis. Support Syst. **57**, 212–223 (2014)
13. Groth, S.S., Siering, M., Gomber, P.: How to enable automated trading engines to cope with news-related liquidity shocks? Extracting signals from unstructured data. Decis. Support Syst. **62**, 32–42 (2014)
14. Hagenau, M., Liebmann, M., Neumann, D.: Automated news reading: stock price prediction based on financial news using context-capturing features. Decis. Support Syst. **55**(3), 685–697 (2013)
15. Han, J., Kamber, M., Pei, J.: Data Mining: Concepts and Techniques. Morgan Kaufmann, San Francisco (2011)

16. Hatzivassiloglou, V., McKeown, K.R.: Predicting the semantic orientation of adjectives. In: Proceedings of the Eighth Conference on European Chapter of the Association for Computational Linguistics, pp. 174–181 (1997)
17. Ma, Z., Sheng, O.R.L., Pant, G.: Discovering company revenue relations from news: a network approach. Decis. Support Syst. **47**(4), 408–414 (2009)
18. Ma, Z., Pant, G., Sheng, O.R.L.: Mining competitor relationships from online news: a network-based approach. Electron. Commence Res. Appl. **10**(4), 418–427 (2011)
19. Manning, C.D., Raghavan, P., Schütze, H.: Introduction to Information Retrieval. Cambridge University Press, New York (2008)
20. Mittermayer, M.-A.: Forecasting Intraday stock price trends with text mining techniques. In: Proceedings of the 37th Annual Hawaii International Conference on System Sciences (2004)
21. Paek, T., Gamon, M., Counts, S., Chickering, D.M., Dhesi, A.: Predicting the importance of newsfeed posts and social network friends. In: Proceedings of the Twenty-Fourth AAAI Conference on Artificial Intelligence, pp. 1419–1424 (2010)
22. Qiu, M., Yang, L., Jiang, J.: Modeling interaction features for debate side clustering. In: CIKM 2013 Proceedings of the 22nd ACM International Conference on Information & Knowledge Management, pp. 873–878 (2013)
23. Schumaker, R.P., Chen, H.: Textual analysis of stock market prediction using breaking financial news: the AZFin text system. ACM Trans. Inf. Syst. **27**(2), 1–19 (2009)
24. Tetlock, P.: Giving content to investor sentiment: the role of media in the stock market. J. Finan. **62**(3), 1139–1168 (2007)
25. Wu, Z., Palmer, M.: Verbs semantics and lexical selection. In: ACL 1994 Proceedings of the 32nd Annual Meeting on Association for Computational Linguistics, pp. 133–138 (1994)
26. Zhai, Z., Liu, B., Zhang, L., Xu, H., Jia, P.: Identifying evaluative sentences in online discussions. Paper presented at the 24th AAAI Conference on Artificial Intelligence (2011)

Acquiring Seasonal/Agricultural Knowledge from Social Media

Hiroshi Uehara[1,2](✉) and Kenichi Yoshida[3](✉)

[1] Graduate School of Systems and Information Engineering,
University of Tsukuba, Ibaraki, Japan
ueharah@nttdocomo.com
[2] NTT DOCOMO, INC., Corporate Sales and Marketing Division, Tokyo, Japan
[3] Graduate School of Business Science, University of Tsukuba, Tokyo, Japan
yoshida@gssm.otsuka.tsukuba.ac.jp

Abstract. Agricultural knowledge depends on seasonally changing conditions such as climate, harmful insects, etc. In this respect, farmers tend to be interested in seasonal knowledge rather than the static principle. To acquire such agricultural knowledge, we propose a method to acquire seasonal knowledge from ongoing posts in the social media. The experimental results shows that the agricultural knowledge can be extracted in the form of chained structures, each of which denotes a set of seasonal knowledge. We also developed a prototype of dialogue robot that provides agricultural knowledge based on the chained structure database. The characteristics of the robot is its ability to reply with seasonally changing knowledge.

1 Introduction

Basic plan for food, agriculture and rural areas [3] is an important policy of Japanese government. IoT technologies are now widely applied to agricultural industry to improve productivity. Most of them take the form of sensor network platform, which comprises of sensors, network, and cloud service to collect the data from sensors, e.g., [5,6,10]. The data from sensors represent growing environment of farm products such as temperature, hours of sunlight and soil conditions. Such data are to be analyzed to provide farmers with forecast of harmful insects and prediction of harvest, etc. Meanwhile, some of the mobile handsets become more suitable for farming. They are equipped with waterproofing, voice recognition and touch panels reactive to work gloves. Farmers can conveniently manipulate the mobile handsets to retrieve knowledge from agricultural sensor network while they are farming.

Although sensor network platforms attract strong attentions as a source of agricultural knowledge, it is not the only source. Some of the social media have a large amount of text data concerning agriculture. They might be another source of agricultural knowledge. Acquisition of such knowledge is especially important for new farmers who Japanese ministry of agriculture, forestry and fisheries tries to help [2]. Although there exists few attempts to acquire agricultural knowledge,

© Springer International Publishing Switzerland 2016
H. Ohwada and K. Yoshida (Eds.): PKAW 2016, LNAI 9806, pp. 129–140, 2016.
DOI: 10.1007/978-3-319-42706-5_10

e.g., [4], they are not so popular yet. Thus, this research focuses on agricultural knowledge acquisition from social media to help their diffusion.

One important characteristics of agricultural knowledge is seasonality. Agricultural knowledge depends on seasonally changing conditions such as climate, harmful insects, hours of sunlight, etc. To acquire such seasonally changing knowledge, we have developed a method to acquire knowledge from ongoing posts in the social media. The characteristics of the method is acquisition of seasonally changing knowledge. Although knowledge acquisition from social media itself is important area of investigation [12], acquisition method itself has a room to investigate. To capture seasonal changes, our approach is different from techniques for extracting conversational knowledge from Twitter such as [9]. Section 2 of this paper first explains the social media we used in this study. Section 3 explains the proposed method. After Sect. 4 reports the experimental results, Sect. 5 explains a prototype of dialogue robot which uses the acquired knowledge to show the practicability of the proposed method. Finally, Sect. 6 concludes our findings.

2   Text Data Concerning Agriculture on the Internet

Lots of books and articles of agriculture exist. Some of these texts have been implemented as the form of knowledge database, which enables farmers to input keyword and easily retrieve related knowledge to the keyword. However, these types of knowledge do not necessarily meet the farmers' needs, partially because farmers use highly conditional knowledge such as appropriate quantity of fertilizer under specific condition of soil, climate, and crop's growth. As such, agricultural knowledge is highly conditional one rather than constant principle written on the books and articles.

Social media, such as Twitter has vast amount of tweets concerning agriculture. For example, the amount of Japanese tweets including the compounds "harvesting rice crops" is estimated to be over 1 million per year. Another example is "Ni-channel" [1], anonymous bulletin board originated in Japan. "Ni-channel" comprise of lots of bulletin boards each of which concerns specific theme of discussion. The examples of agricultural bulletin boards are shown in Table 1 with

Table 1. Bulletin boards of "Ni-channel" concerning agriculture

Title of the bulletin board	Amount of posts
Rice farmers	45,000
Mowing machines	60,000
Dairy farmers	30,000
Enter into farming	80,000
Live stock farming	49,000
Total	264,000

the amounts of posts in this few years. In this research, we try to acquire the agricultural knowledge from the bulletin board of rice farmers.

3 Acquiring Agricultural Knowledge from "Ni-channel"

This section proposes a method to acquire agricultural knowledge from a social media so called "Ni-channel" [1]. The characteristics of the proposed method are:

- It is designed to acquire important topics from "Ni-channel". Such topics depend on seasonally changing conditions such as climate, harmful insects, etc.
- The importance of the acquired topic at some season is represented as peaks in line chart (See Sect. 3.1).
- The proposed method extracts words as important topics. Related words for the same topics are extracted by analyzing the co-occurrence of words in the same posts (See Sect. 3.2).
- Although the importance of topics changes with season changes, the proposed method traces such changes based on the peak in line chart. It stores the extracted trace in the form of chained structure (See Sect. 3.3).
- The proposed method also extracts important compounds from the posts. (See Sect. 3.4).

Here the proposed method is based on the method proposed in [13,14]. The extraction of important compounds and the generation of the chained structure which represents topic change are the newly developed functionality.

3.1 Detecting the Importance of Topics

According to prior related research [13,14], the frequency of words representing topic in each bulletin board tends to increase in accordance with the importance of the topic in a given period. In other words, the word frequency represents the importance of the topics in some time period. Meanwhile, the words other than the ones of interest show sporadic and random appearance patterns.

Making use of this characteristics, we have proposed the algorithm to quantify the degree of participants' interest in [13]. Figure 1 shows the example. In Fig. 1, each box represents a post on "Ni-channel". The word "harvesting" appears one after the other in the period N while it appears every other post in the period N+1. In this case, the importance of the topic represented by "harvesting" in period N is higher than the one in period N+1. In this manner, shorter consecutive intervals mean higher importance of the topic.

Our algorithm has the threshold of the reappearing interval of words to cut off less important topics. For example, the threshold of reappearing interval setting to 4 result in sporadic line chart as Fig. 2. Each peak in line chart represents life cycle of topic, that is, zone surrounded by both edge of each peak represents

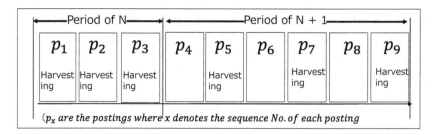

Fig. 1. Measuring the importance of the posts

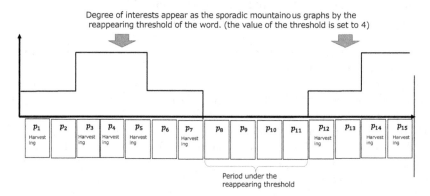

Fig. 2. Peaks in line chart which express the importance of topics

duration in which the topics holds its importance. Additionally in order to measure confidence for the importance of each topic, Repetition count of the interval of the word is adopted as another threshold. For example in the period of N+1 of Fig. 1, repeating just twice of "Harvesting" on p_5, p_7 at the interval of 1 might be insufficient to believe the topic represented by "Harvesting" has the importance equivalent to the interval 1. Alternatively 3 times repetition, i.e., appearing on p_5, p_7 and p_9 at the interval might be sufficient to believe it. In this case, the threshold for the confidence of the interval is set to 3.

3.2   Finding Related Words for Topics

Applying the algorithm above often generates co-occurring peaks in line chart. Some of them might belong to the same topic and the others might not. In case of social media, posts implying the same topic can be connected by chain structures which represent users' interactions. Generally the chained structures are explicitly described by the replying symbols such as "@" or ">". However, Ni-channel does not necessarily have such replying symbols so that alternative clues should be considered to classify co-occurring peaks into relevant topic categories.

Posts implying the same topic tend to share the related words. Figure 3 shows the example of posts that constitute co-occurring peaks in line charts. The post

Fig. 3. Finding related words for topics

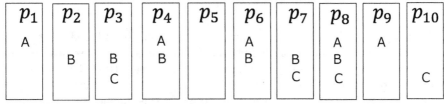

Threshold of co-occurrence repetition is set to 3
(Threshold of the reappearing interval of words is set to 4)

Fig. 4. Detecting the topics changes from co-occurring peaks

p_1 and p_2 include the same word "harvesting", and the post p_2 and p_4 include the same word "golden". Similarly, the same words "rice" and "fields" co-appear in both of p_4 and p_5. In this case, we consider "golden" being inspired by "harvesting". We also consider "rice" and "fields" being inspired by "golden".

Here, only one co-occurrence of "harvesting" and "golden" in p_2 is not sufficient to believe that "golden" is inspired by "harvest". Confidence measure should be introduced so that "harvest" and "golden" likely to belong to the same topic of interest. We adopt repeat count of co-occurrence to measure the confidence. This confidence measure is similar to the threshold introduced in Subsect. 3.1, where the measure is repeat count of re-appearing interval of the word. In Fig. 4, threshold of co-occurrence repetition is set to 3. Here, two words co-occurring less than 3 times are not recognized as the same topic of interest. Based on this rule, word A and B are recognized as the same topic. B and C are also the same case. However, A and C are not. Consequently, the threshold defines the overlapping degree of co-occurring peaks.

3.3 Detecting Users' Interacting Sequence from the Words of the Same Topic

Although the rule above clarifies the correspondence between co-occurring peaks and the topic, still it is not clear that B is inspired by A or vice versa. Again Fig. 4 shows that left edge of peak A start from post 1, preceding to the one of peak B. Similarly the peak B precedes to C. The direction is determined based on this sequence context of peaks, the peak having preceding left edge is recognized as the parent of peak that has following left edge.

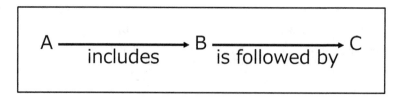

Fig. 5. Extracted chain structure

Figure 5 shows the chained structure extracted from Fig. 4 based on the rule above. The arrows indicate the direction of the words, i.e., B is inspired by A, then C is inspired by B, etc. The chain also reveals that C is indirectly connected to A via B. The sequence from A to C might imply the direction of users' interactions concerning the same topic. In this manner such a chained structure reveals unique sequence of users' interactions sharing the same topic. The annotation to each arrow in Fig. 5 represents overlapping pattern of co-occurring peaks. Both edges of the peak A are corresponding to the posts 1 and 9 which include both edges of peak B inside (posts 2 and 8). In this case, the relevant arrow is annotated by "A includes B" (hereafter "including relation"). Meanwhile right edge of the peak C is located after the right edge of the peak B while both of the peaks follow overlapping rule above. In this case, the relevant arrow is annotated by "B is followed by C" (hereafter "following relation"). Following relations might imply developments of users' interaction. C is inspired by B initially, and might be developed into the relating topic other than the topic including A and B.

3.4   Detecting Compounds from the Words of the Same Topic

In Fig. 6, "Typhoon" and "George" in the posts 1, 2 and 4 implies topic of the same interest. Moreover they appear consecutively in each post. The degrees of interest to both words show the same pattern of fluctuation. In such case, words are considered to be part of same compounds. In order to assure confidence, threshold similar to the repeated count of co-occurrence explained above is applied. If the threshold is set to 3, posts 1 and 2 are not sufficient to recognize that "typhoon" and "George" should be paired. Posts 1, 2 and 4 are sufficient for the recognition.

Fig. 6. Detecting words to be recognized as one compound

4   Experimental Results

In this section, we first report the extracted compounds with their importance at some time period (Sect. 4.1). Then we explain the sequence of topic change in accordance with season change (Sect. 4.2).

4.1   Analyzing Peaks of Users' Interests and Extracting Compounds

We analyzed data of rice farmers' bulletin board of "Ni-channel" in 2014 using the method explained in Sect. 3. Summary of data is shown in Table 1. The number of posts during the period is 6500. Figure 7 shows the results of applying the algorithm explained in 3.1. X axis is the period of data. Y axis represents the degree of importance of each word appeared in the bulletin board. Approximately 400 words are detected as the topic of interest.

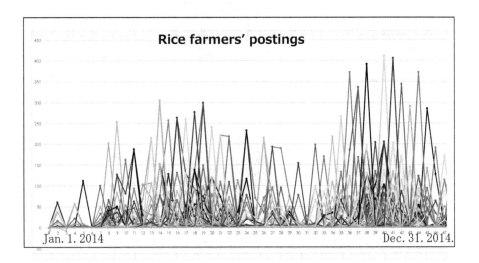

Fig. 7. Chart representing the degree of interests

Figure 8 shows part of data in Fig. 7. In Fig. 8, (a) shows the results concerning harmful insects. Each word represents the harmful insect of rice plants. Each period of peaks coincides with the period when corresponding insect tend to make problems. By applying the method described in Sect. 3.4, a set of words which contains "jumbo", "mud", and "snails", is recognized as one compound. Here, "Mud snails" are beneficial insects in general. However "Jumbo mud snails" are the exception. It arises at the period for setting out rice plants. Since "Jumbo mud snail" eats seedlings, it causes serious problem at the period of setting out rice plants. This period coincides with the peak in the chart. These results show that the proposed method can detect important compound "Jumbo mud snail" as harmful insects of rice plants.

Similarly, (b) is the extraction concerning farm works. Again words concerning rice plants' farm works, such as "deep watering", appear in accordance with each farm working period in paddy fields. Another example of detecting compound is "solar" and "drying". Rice in the husk are dried under the sun right after harvesting.

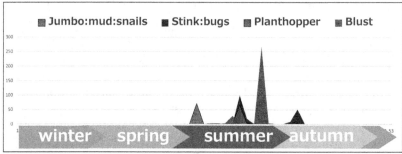

(a) Graphs of the words concerning harmful insects

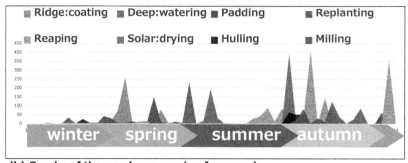

(b) Graphs of the words concerning farm works

Fig. 8. Samples of the peaks in line chart

4.2 Analyzing Users' Interacting Sequence from Co-occurring Peaks

The topics on the social media changes in accordance with the season changes. The proposed method classify peaks in Fig. 7 into topics of the same interests. Then the chained structures are generated from co-occurring peaks to express users' interacting sequence concerning the topic. Thus, each chained structure represents the topic change on the social media in accordance with season change.

Figure 9 is the one of the extracted chain structures. As a whole, the chain expresses the topic concerning padding, i.e., the farm work before setting out the rice plants. Normally in order to do paddings, farmers stir water into the soil with special attachment to tractor called rotor. Tractors are also used for coating rice paddy's ridges with mud to prevent water leakage. Herbicides are

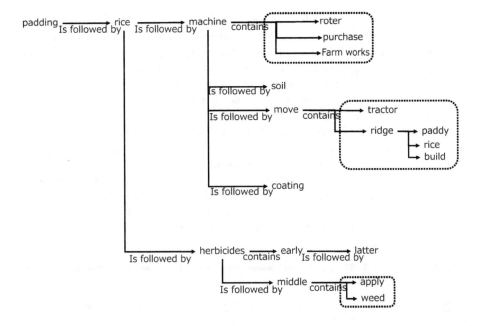

Fig. 9. Chain structure of the words belonging to the same topic

applied after the padding, and are classified into three types depending on the timing of application, early periods, middle periods, and latter periods.

The chained structure in Fig. 9 is likely to follow the farm work for the padding. Two kinds of chains are branched off from the root annotated by "padding" and "rice". One branch consists of the chains concerning the machines for paddings, which is again branched off to two kinds of machines for different usages. In this chains "machine" is an umbrella of "rotor", "purchase", "farm work" implying those three words appear in the same context of users interaction. "Move" is another umbrella of "tractor" and "ridge", under which "paddy", "rice", "build" also belong to the same context. The other branch consists of the chains concerning herbicides. Users' interactions likely focus on three types of herbicides.

Note that there exist studies which try to extract sequence of dialogue from social media, e.g., [8,9,11], the length of dialogue extracted by these conventional studies tends to be short (topically 2-3 tweets). The long/tree-like chained structure shown in Fig. 9 is an unique characteristics of our proposed method.

5   Prototype of Dialogue Robot

The chained structure represents the knowledge of the interactive sequence in the same topic of interest. Users seem to interact their knowledge in line with this structure. The words of interests are the symbols of the knowledge. Based on this assumption, we manually extract the phrases including the words of interests that are considered to be useful for agricultural knowledge. Then, we implement them as a form of database in which the phrases of agricultural knowledge are linked with each other in accordance with the chain structure. This chained data structure can be considered as the dialogue sequence of knowledge.

Making use of this data structure, we make a prototype dialogue robot that automatically provides the agricultural knowledge stored in the chain structure. It is a prototype that intends to make a new service similar to [7]. Figure 10 illustrates how the prototype works. If users input a dialogue phrase in the text form at the bottom, the system search the database with the input phrase, and fetch one of the phrases linked to the input phrase. The robot replies to the users' dialogue displaying the fetched phrase. Because multiple phrases of knowledge are linked to the parent phrase (i.e., input phrase), the system select one of them randomly. Therefore, every time when users input the same dialogue phrase, different replies are created. (a)-1, and (a)-2 in Fig. 10 are the examples.

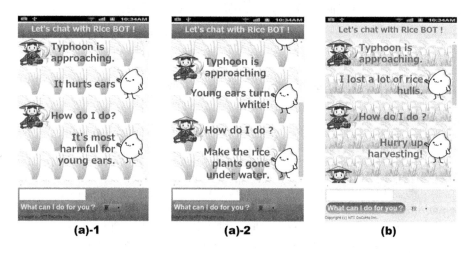

Fig. 10. Prototype of dialogue robot

The robot also takes seasonal information into consideration to make reply. Both (a) and (b) in Fig. 10 are the dialogue concerning typhoon. (a) is the example of the middle of summer when rice plants come into ears. (b) is the example of the beginning of autumn when they are harvested. Linked knowledge of both cases are different with each other. They are created from the different co-occurring peaks of users' interests. Robot's replies in (a) and (b) reflect seasonal difference of the linked knowledge.

6 Conclusion

In this research we tried to acquire the agricultural knowledge from the posts of Ni-channel. Applying the algorithm for detecting the peaks of users' interests revealed that each peak of extracted topic coincident with the period of seasonal knowledge. The characteristics of the proposed method are:

- The proposed method is designed to acquire important topics from social media. Such topics depend on seasonally changing conditions such as climate, harmful insects, hours of sunlight, etc.

  The importance of the acquired topic at some season is represented as peaks in line chart.
- The proposed method also extracts important compounds from the posts by analyzing the co-occurrence of words.
- The importance of topics changes with season changes. The proposed method traces such changes based on co-occurring peaks in line chart. It stores the extracted trace in the form of chained structure.

  The long/tree-like chained structure which stores the sequence of dialogue is an unique characteristics of the proposed method.

The experimental results show the functionality of the proposed method. The extracted knowledge through the experiments are stored in the chained structure database, and used to realize a prototype dialogue robot which is able to provide seasonally changing agricultural knowledge.

Acknowledgment. This work was supported in part by JSPS KAKENHI Grant Number 25280114.

References

1. http://2ch.net/
2. Ministry of Agriculture, Forestry and Fisheries. http://www.maff.go.jp/j/new_farmer/
3. Ministry of Agriculture, Forestry and Fisheries: Summary of the Basic Plan for Food, Agriculture and Rural Areas. http://www.maff.go.jp/e/pdf/basic_plan_2015.pdf
4. http://lib.ruralnet.or.jp/nrpd/
5. http://www.itmedia.co.jp/promobile/articles/1207/11/news121.html
6. http://news.mynavi.jp/news/2013/07/19/014/
7. Docomo, N.: https://www.nttdocomo.co.jp/service/information/shabette_concier/
8. Higashinaka, R., Kawamae, N., Sadamitsu, K., Minami, Y., Meguro, T., Dohsaka, K., Inagaki, H.: Building a conversational model from two-tweets. In: 2011 IEEE Workshop on Automatic Speech Recognition and Understanding (ASRU), pp. 330–335. IEEE (2011)
9. Higashinaka, R., Kobayashi, N., Hirano, T., Miyazaki, C., Meguro, T., Makino, T., Matsuo, Y.: Syntactic filtering and content-based retrieval of twitter sentences for the generation of system utterances in dialogue systems. In: Proceedings of the IWSDS, pp. 113–123 (2014)
10. Hirafuji, M.: Application of sensor network in agriculture. Proc. TMS **2009**(1), 22–28 (2009)

11. Inaba, M., Kamizono, S., Takahashi, K.: Candidate utterance acquisition method for non-task-oriented dialogue systems from twitter. Trans. Jpn. Soc. Artif. Intell. **29**, 21–31 (2014)
12. Kim, Y.S., Kang, B.H., Richards, D. (eds.): Knowledge Management and Acquisition for Smart Systems and Services. Springer, Heidelberg (2014)
13. Uehara, H., Yoshida, K.: Annotating tv drama based on viewer dialogue - analysis of viewers' attention generated on an internet bulletin board. In: Proceedings of the SAINT, pp. 334–340 (2005)
14. Uehara, H., Yoshida, K.: Acquiring marketing knowledge from internet bulletin boards. In: Richards, D., Kang, B.-H. (eds.) PKAW 2008. LNCS, vol. 5465, pp. 173–182. Springer, Heidelberg (2009)

Amalgamating Social Media Data and Movie Recommendation

Maria R. Lee[1(✉)], Tsung Teng Chen[2], and Ying Shun Cai[1]

[1] Department of Information Technology and Management,
Shih Chien University, Taipei, Taiwan
Maria.lee@g2.usc.edu.tw
[2] Graduate Institute of Information Management,
National Taipei University, New Taipei City, Taiwan
timchen.ntpu@msa.hinet.net

Abstract. Recommender systems (RSs) have become very common recently. However, RS techniques need large amounts of user and product data, which hinders RS usage for businesses with insufficient data. The RS cold-start problem may be mitigated by leveraging external data sources. We demonstrate the feasibility of solving the cold-start problem by implementing a hybrid RS that integrates the Facebook Fan Page data and the genre-classifications data from Yahoo! Movies. Our study amalgamates social media data and machine learning to build a hybrid-filtering RS. We also compared our system with three existing movie RSs—those used by Netflix, YouTube, and Amazon. Within the framework of a hybrid-filtering RS, content-based filtering was used to extract data from Yahoo! Movies and Facebook Fan Pages. The proposed RS overcame the cold-start problem and achieved a satisfactory level of accuracy.

Keywords: Recommender system · Social media data · Machine learning · Netflix · Youtube · Amazon

1 Introduction

With the increasing ubiquity of the Internet, social networking sites (SNSs) have also become increasingly sophisticated (Sarwar et al. 2000; Schafer et al. 2001; Lee et al. 2002; Kaplan and Haenlein 2010; Liu et al. 2013; Lu et al. 2014). Moreover, Facebook provides an application program interface (API) for retrieving user and Fan Pages data. Therefore, there is a promising opportunity for value creation if these data can be utilized by businesses and be collated with industry knowledge and analysis.

Recommender systems (RSs) play an indispensable role in e-commerce and increase the profits of businesses intending to sell merchandise (Hirji 1999; Jannach et al. 2010; Bobadilla et al. 2013). However, consumers demand different products, and users who are not familiar with the product will often make poor purchasing decisions. Hence, businesses need to determine the consumers' preferences and recommend suitable products.

The most commonly applied RSs in the mainstream use collaborative filtering and content-based filtering (Goldberg et al. 1992; Burke 2002; Ahn 2008; Al-Shamri and

© Springer International Publishing Switzerland 2016
H. Ohwada and K. Yoshida (Eds.): PKAW 2016, LNAI 9806, pp. 141–152, 2016.
DOI: 10.1007/978-3-319-42706-5_11

Bharadwaj 2008). However, the data for both collaborative filtering and content-based filtering are obtained from the businesses' internal members and product databases (Said et al. 2011). Hence, because of the insufficient volume of product types and member data, the implementation of the aforementioned RSs proves to be challenging for small and medium-sized enterprises (SMEs) or creative industries, and the accuracy of the recommendations is below expectations. In view of this problem, this study combined content-based filtering and collaborative filtering to develop a movie RS which incorporates data from SNSs.

Currently, three major film and television service providers are available—Netflix, YouTube, and Amazon—all of which employ RSs (Szomszor et al. 2007; Debnath et al. 2008; Lekakos and Caravelas 2008, Korenl et al. 2009; McSherry and Mironov 2009; Adomavicius et al. 2010). This is a key service offered by these firms to provide more precise and individualized recommendations for each user. Such a service is helpful to users and may spark their interest, which brings in more profit for the businesses. However, the types of content provided by the three companies differ from each other. Netflix is mainly a video-streaming service for TV series or movies; YouTube is a video-sharing website that allows streaming of short video clips; Amazon is an e-commerce website that supplies digital goods. This article will explore the three major movie RSs and discuss their pros and cons. Subsequently, we will introduce our proposed system, and compare the differences between the four systems.

Creating an RS for SMEs that can fulfill client demands and give accurate recommendations will certainly enhance their competitiveness. Therefore, we developed an RS that does not require building an internal membership database. We demonstrate the feasibility of implementing a hybrid RS that integrates the Facebook Fan Page data and the genre-classifications data from Yahoo! Movies.

RSs for the premiere movies still employ traditional methods, which use texts or videos in Taiwan marketing practices. Before deciding whether to watch a movie, one must spend considerable time searching related websites (e.g., Yahoo! Movies, PTT Movie Board, and @ Movies) for reviews posted by fellow users. However, each person has their own subjective perception, and reviews on the same movie could be very different, which is not very helpful for the user. Therefore, building an SNS-based movie RS could facilitate public decision making, thereby reducing the time spent searching for movie information and increasing the satisfaction of movie viewers (Cai and Lee 2015). This study proposed an SNS-based, machine-learning, hybrid-filtering RS. Within the framework of a hybrid-filtering RS, content-based filtering was applied to extract data from Yahoo! Movies and Facebook Fan Pages.

2 Literature Review

Facebook includes three main features: the Facebook Wall on the personal page, Groups, and Fan Pages (Lu et al. 2014). Fan Pages are completely open to the public with no limit on the number of users who can join, and users are free to choose whether to join a Fan Page. Facebook also provides the corresponding API for users and Fan Page data; thus, allowing third parties to develop Facebook-based application services while also providing developers and researchers with the means to retrieve Facebook data. This has led to the blossoming of Facebook third-party applications and the increasingly widespread application of value-added services, thereby consistently boosting the number of users and enabling Facebook to become the world's fastest-growing website.

An RS's main aim is to provide recommendations to users. It is an intelligent network system that assists users to search for items that they are interested in or that fulfill their desires' work, from among the wealth of products, services, and information available on the Internet. An RSs survey conducted by Bobadilla (2013) shows that recent research has been predominantly content-based, collaborative, demographic, and hybrid filtered. However, because demographic filtering is computed using general statistical methods and, hence, can be directly integrated with the content-based and collaborative filtering RSs, it is not discussed in this article.

A content-based filtering RS employs information-extraction and information-filtering techniques. It is a text-based system that uses keywords or tags to filter meaningful product descriptions, and refers to the users' browsing or purchase histories to establish related features between the users and the products. Machine-learning models are needed to identify and learn about users' interests or behavior patterns. Recommendations are further based on the degree of similarity with the training model (Balabanović and Shoham 1997).

Collaborative-filtering RSs process a vast amount of data through e-mail classification. Collaborative filtering mainly involves using the ratings on products and items from similar users within the same group as the basis for providing recommendations to other members (Sarwar et al. 2001; Davidson et al. 2010).

Hybrid filtering was primarily derived from the need to resolve the disadvantages of individual RSs. It uses the advantages of two or even three techniques to compensate for the limitations of the original techniques; the most common combination is collaborative and content-based filtering. This combination enhances the accuracy and solves the cold-start problem caused by new users and new products in content-based and collaborative filtering.

In terms of the recommended products, Netflix mainly recommends the online streaming of TV series or movies; YouTube recommends its platform for users to upload short video clips; and Amazon recommends multiple categories of digital goods (Zimmerman et al. 2004; Amatriain and Basilico 2012). For the recommendation techniques, Netflix uses a hybrid recommender system (collaborative filtering and content-based filtering) (Gower 2014); YouTube utilizes content-based filtering and computes the similarity of each clip (Davidson et al. 2010); and Amazon employs collaborative filtering and uses its products as the basis for recommendation (Linden et al. 2003). All three systems must establish their own member system before they can provide recommendations and are unable to provide accurate recommendations in the beginning, due to the lack of historical datasets on new users and products.

Computation Time. Although Netflix uses rapid matrix operations, the vast data volume that has grown over time implies that the computation will need to be replaced by parallel or distributed computing. YouTube calculates the similarity between each video clip, which requires pre-processing to provide rapid recommendations when users are using the service. Amazon uses cosine similarity combined with their algorithms to significantly reduce the computation time. All three systems have problems in terms of data sparsity, and their quality depends on the historical datasets, but they do not require a knowledge of the domain to obtain the recommendation results.

Complexity. Netflix is simple, as it converts the products and users into matrices. YouTube uses the similarity computation of video clips uploaded by users within the last 24 h, and those clicked by users. Due to the sheer number of users on YouTube, the similarity computation of each clip is extremely complicated. Since Amazon has an extensive products portfolio, the data required by the RS are solely based on their own database for the computation of product similarity, which has been automated and is relatively simple. Table 1 shows a summary of the three recommendation systems.

Table 1. Comparison of recommending techniques

	Netflix	Youtube	Amazon
Recommended item	Video & Movie streaming	User video	Digital items (Movie & Music..)
Recommended techniques	A hybrid recommender system	Content-based filtering	Hybrid-based filtering (product-oriented)
Need membership	Yes	Yes	Yes
New user problem?	Yes	Yes	Yes
Computation time	Slightly fast	Slow	Fast
Data sparsity	Depends on the historical datasets	Depends on the historical datasets	Depends on the historical datasets
Domain knowledge required	No	No	No
Complexity	Simple	Similarity computation required	Simple

3   Research Method

This study combined external data from Facebook and Yahoo! Movies to create an RS, using premiere movies in Taiwan as the recommended items, and employing PHP to extract data. Subsequently, in data pre-processing, R was used to convert the data into a format compatible with collaborative-filtering techniques. Thus, a model was created for each movie genre using the collaborative-filtering technique in the R language environment to construct a hybrid-filtering RS (Ihaka and Gentleman 1996).

Once the prototype was completed, historical data were introduced into the system for cross-validation, to evaluate the accuracy level, as well as to adjust the parameters of the system; e.g., the number of Fan Pages, screening for historical data testing, dimensions to be reduced in singular-value decomposition (SVD), and selection of regression models. After adjustments were made, the accuracy level showed some improvements and the prototype could be used to conduct experiments. Thereafter, an RS was created using PHP and R, based on the API provided by Facebook, followed by a survey questionnaire. After processing and analyzing the returned questionnaires, the data were used to verify the effectiveness of the RS, as well as evaluate the differences between past well-known RSs and our proposed RS.

This study proposed an SNS-based, machine-learning, hybrid-filtering RS (Fig. 1). Within the framework of a hybrid-filtering RS, content-based filtering was applied for data extraction, including data crawling of Yahoo! Movies and Facebook Fan Pages, as well as data extraction of user information.

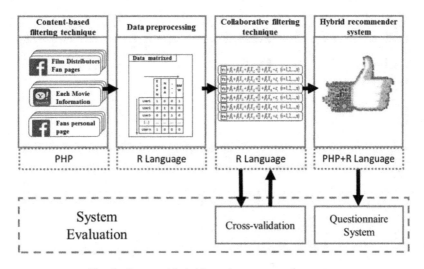

Fig. 1. Proposed hybrid movie recommender system

At the pre-processing stage, data conversion involved transforming data into matrix format, which enabled processing for collaborative filtering (Koren et al. 2009). Fan Pages were pre-sorted and used as related factors during the determination of recommendations. SVD was applied for dimension reduction to enable more-efficient computations. Finally, the data were imported into the collaborative filtering RS, and a machine-learning-based multinomial logistic regression model was constructed. The accuracy of the model was evaluated using the test dataset, and served as a reference for subsequent research.

4 Proposed System Demonstration

Figure 2 shows the proposed system architecture. Content-based filtering RSs stem mainly from two fields: information extraction and information filtering. This study applied the same principles: first, extracting individual movie attributes from the Facebook Fan Pages, Facebook personal pages, and Yahoo! Movies, followed by information filtering utilizing posts on Facebook Fan Pages. The primary aim of this study is to construct an RS for premiere movies, which links the movies with the content of Facebook Fan Page posts, thereby identifying which movies were "Liked" by each Facebook user.

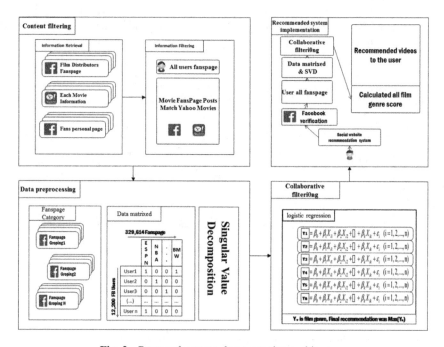

Fig. 2. Proposed system demonstration architecture

A Fan Page post might mention the movie title keywords; however, some posts might not be an exact match for the movie titles in Yahoo! Movies, because of differences in the input methods, e.g., differences in full-width and half-width forms in Chinese characters, and differences in punctuation. To enhance the accuracy level, punctuation was removed from each post, and then a PHP function was applied to implement string matching.

If a movie match was found after the program execution, the movie title and the ID of the Fan Page post were stored in the database. If no matches were found, the search continued sequentially until the end. The collated data from Facebook Fan Pages and Yahoo! Movies are shown in Table 2.

Table 2. Collated data from Facebook fan pages and Yahoo! Movies (Cai and Lee 2015)

ID	Film Name	Film Genre	Posts	Likes	ID	Film Name	Film Genre	Posts	Likes
1	Annabelle	Mystery	116	151,832	12	But Always	Love	69	24,512
2	The Maze Runner	Mystery	118	328,522	13	Turning Tide	Drama	36	7,001
3	The Frogville	Comedy	223	34,432	14	Blind	Drama	71	35,725
4	This Is Where I Leave You	Drama	93	45,537	15	Words and Pictures	Love	44	2,906
5	Paradise In Service	Drama	220	124,142	16	Deliver Us from Evil	Mystery	72	105,850
6	Lucy	Action	135	473,283	17	The Expendables 3	Action	155	59,850
7	The World of Kanako	Mystery	44	46,081	18	Before I Go to Sleep	Mystery	139	45,352
8	As Above, So Below	Terror	55	67,312	19	Chef	Inspiration	165	37,464
9	The Boxtrolls	Animation	60	59,994	20	The Purge: Anarchy	Mystery	35	35,534
10	The Sacrament	Drama	26	21,579	21	Partners in Crime	Mystery	237	144,450
11	Sex Tape	Comedy	83	74,305	22	Wood Job!	Action	106	6,446

The main aims of the data pre-processing stage include: (1) grouping the Fan Pages to explain the influential factors after analyzing the recommendation results; (2) data conversion to matrix format so that the data can be read and processed by the collaborative-filtering model; and (3) data-dimension reduction. Up to 329,614 items were collected from Facebook Fan Pages, which involved 12,200 users. Thus, to optimize the computation efficiency while also preventing the loss of data, SVD was applied to reduce the dimensionality and to extract the eigenvalues of the original data.

Collaborative filtering involves searching whether past users have eigenvalues similar to the recipient of recommendations, and is based on the fact that similar users might share similar preferences. As for the collaborative-filtering component, this study used a multinomial logistic regression technique for investigation. This is a supervised machine-learning algorithm that requires the input of eigenvalues and target values for training. The input values were from the user Fan Page matrix obtained after SVD processing, and the target values were from whether a Facebook user Liked a particular movie. If yes, the target value was denoted by 1, and if not, it was denoted by 0. The parameter values of each movie could be obtained after training. In the multinomial regression of collaborative filtering, a linear regression model was established, whereby the dependent variable **y** was the movie genre, which included suspense/thriller, comedy, action, drama, horror, animation, romance, adventure, and inspirational.

The target values for these nine items depended on whether a user was interested in this genre, denoted by 1 if interested, and 0 if not. Training was then conducted through a machine-learning model. The equation model is as follows:

$$Y_i = W_{g0} + W_{g1}X_{ug1} + W_{g2}X_{ug2} + \ldots\ldots + W_{g99}X_{ug99} + W_{g100}X_{ug100} \qquad (1)$$

Y is the movie genre; W_{g0}: the constant value; W_{g1} to W_{g100}: movie genre independent variables (weights); X_{ug1} to X_{ug100}: user Fan Page independent variables (parameter).

The algorithm amalgamates social media data and machine learning to create a hybrid-filtering RS. SVD was used to reduce the dimensionality, and multi-logistic regression was used to train and establish the *Film genre* module.

1	**Algorithm for movie recommender**
2	**For each** *Post in Fanspage* **in** *Film Genre*
3	**For each** *Post Like user* **in** *Post in Fanspage*
4	collect *Post Like user 's Like Fanspage*
5	set the *Post Like user - Fanspage* Matrix, M_1
6	use Singular Value Decomposition to dimensionality reduction M_1
7	get I_1
8	
9	**For each** I_1 **in** *Film genre*
10	Use Muti-logistic regression training and establishing of
11	*Film genre module.*
12	
13	**IF** new user to be recommended
14	**Then**
15	collect *Post Like new user 's Like Fanspage*
16	set the *Post Like new user - Fanspage* Matrix, M_2
17	use Singular Value Decomposition to dimensionality reduction M_2,
18	get I_2
19	
20	**For each** *Film genre module*
21	put I_2 to *Film Genre module*
22	get the each *Film genre score*
23	
24	**For each** *Film genre score*
25	**IF** *score* is the highest
26	**Then** Recommend that *Film genre*

5 System Evaluation

The system evaluation of our proposed RS was divided into two parts. The first part was cross-validation to investigate the accuracy of the logistic regression model for each movie genre. K = 10 indicated that 90 % of the overall fan population was used as the training dataset, while the remaining 10 % was used as the test sample. The accuracy for each genre is shown in Fig. 3. After validation, the overall mean judgment accuracy was more than 65 %, but lower than 50 % for the suspense/thriller genre.

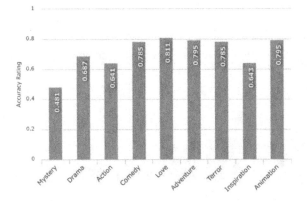

Fig. 3. Results of cross validation evaluation (Cai and Lee 2015)

The second part was the online system questionnaire, and our research subjects were selected from the Facebook user population. The online system-questionnaire evaluation process is shown in Fig. 4. During the testing period, eight premiere movies were recommended, each in the 50th, 51st, and 52nd weeks, giving recommendations for a total of 24 movies. The movie genres included 10 dramas, 4 actions, 3 comedies, 3 horrors, 2 romances, 1 adventure, and 1 animation.

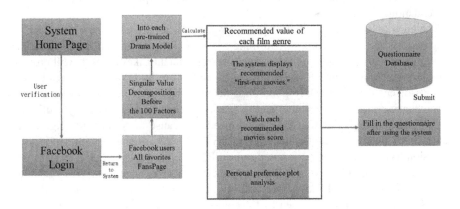

Fig. 4. On-line system questionnaire evaluation process

The questionnaire asked users to rate the individual items recommended by our RS and their overall perception. For all experiences, the maximum score was 10 points, and the minimum was 1 point. The overall mean score for all users was 8.52 points.

The overall results of the online system questionnaire evaluation are shown in Fig. 5. Forty-eight out of 102 people gave 10 points (highest score), 13 people gave 9 points, 15 people gave 8 points, 12 people gave 7 points, 5 people gave 6 points, 6 people gave 5 points, 2 people gave 3 points, and 1 person gave 2 points. No one gave the lowest score (1 point).

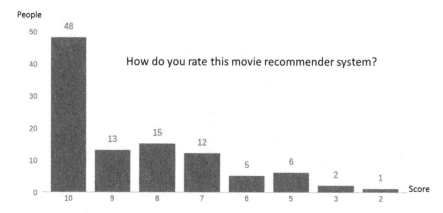

Fig. 5. Results of on-line system questionnaire evaluation

6   Discussion and Conclusion

With regard to the RS techniques applied in our proposed system, content-based and collaborative filtering were combined to form a hybrid-filtering RS. No initial member database is required since we can utilize the membership information from Facebook. To avoid the problems caused by new users, the user's Fan Page combined with SVD could be used to compute the scores of the recommended products, and provide recommendations for the users.

The creation of Facebook Fan Pages is required for all products, new and old. The computation time only involves performing SVD on all Likes by Fan Page users. Due to the data-sparsity problem, only Fan Pages with sufficient Likes by users will provide accurate predictions. Our RS is extremely dependent on the knowledge-classification methods of each domain. Only good classifications will allow us to achieve a more comprehensive model for training, and to reach a certain level of accuracy. Sufficient background knowledge and techniques are needed because of complexity; e.g., using the Facebook data concatenation method and SVD, as well as the application of logistic regression to construct this SNS-based hybrid RS.

This study developed a movie RS combined with SNS data, and used the Facebook historical dataset for validation. We also compared the three major movie RSs to address the accuracy issues caused by the cold-start problem; thus, enabling our RS to achieve a satisfactory level of accuracy immediately after it was introduced online, while also solving the new-user problem. Furthermore, our RS also achieved rapid computation of the customized recommendation items.

Text analysis was involved in the recommendation process, and was based mainly on traditional Chinese; hence, experiments and investigations in other languages were not performed. Since premiere movies are limited by the duration of their release in cinemas, the products were renewed every week; hence, movies by genre were used to supplement the recommendation mechanism in this study.

Facebook Graph API version 2.5 was used to construct the RS in this study. If Facebook updates or terminates the provision of its user information, e.g., user ID, all Fan Pages Liked by users, Fan Page information, and the number of Likes on Fan Pages, our method will be unable to run. Furthermore, if users using this RS do not have a Facebook account or if the number of Fan Pages Liked is less than 300, this RS is not appropriate.

In terms of the machine-learning algorithm, future research should consider using multidimensional algorithms, such as neural networks or fuzzy logic, to replace multinomial logistic regression to compare the different algorithms and make modifications to enhance the accuracy.

Acknowledgement. The article is partially supported by MOST-104-2410-H-158-008 and USC-104-05-04001.

References

Adomavicius, G., Tuzhilin, A., Berkovsky, S., De Luca, E.W., Said, A.: Context-aware recommender systems: research workshop and movie recommendation challenge. In: ACM RecSys2010, pp. 26–30 (2010)

Ahn, H.J.: A new similarity measure for collaborative filtering to alleviate the new user cold-starting problem. Inf. Sci. **178**(1), 37–51 (2008)

Al-Shamri, M.Y.H., Bharadwaj, K.K.: Fuzzy-genetic approach to recommender systems based on a novel hybrid user model. Expert Syst. Appl. **35**(3), 1386–1399 (2008)

Amatriain, X., Basilico, J.: Netflix recommendations: beyond the 5 stars (part 1). Netflix Tech Blog (2012)

Balabanović, M., Shoham, Y.: Fab: content-based, collaborative recommendation. Commun. ACM **40**(3), 66–72 (1997)

Bobadilla, J., Ortega, F., Hernando, A., Gutiérrez, A.: Recommender systems survey. Knowl.-Based Syst. **46**, 109–132 (2013)

Burke, R.: Hybrid recommender systems: Survey and experiments. User Model. User-Adap. Inter. **12**(4), 331–370 (2002)

Cai, Y.S., Lee, M.R.: Research on social network film genre recommender system. In: The 21th Cross Strait Conference on Information Management development and Strategy, Macao, China (2015)

Davidson, J., Liebald, B., Liu, J., Nandy, P., Van Vleet, T., Gargi, U., Gupta, S., He, Y., Lambert, M., Livingston, B.: The YouTube video recommendation system. In: Proceedings of the Fourth ACM Conference on Recommender systems, pp. 293–296. ACM, New York (2010)

Debnath, S., Ganguly, N., Mitra, P.: Feature weighting in content based recommendation system using social network analysis. In: Proceedings of the 17th International Conference on World Wide Web, pp. 1041–1042. ACM, New York (2008)

Goldberg, D., Nichols, D., Oki, B.M., Terry, D.: Using collaborative filtering to weave an information tapestry. Commun. ACM **35**(12), 61–70 (1992)

Gower, S.: Netflix Prize and SVD (2014). http://buzzard.ups.edu/courses/2014spring/420projects/math420-UPS-spring-2014-gower-netflix-SVD.pdf. Accessed 8 Jan 2016

Hirji, K.K.: Discovering data mining: From concept to implementation. ACM SIGKDD Explor. Newsl. **1**(1), 44–45 (1999)

Ihaka, R., Gentleman, R.: R: a language for data analysis and graphics. J. Comput. Graph. Stat. **5**(3), 299–314 (1996)

Jannach, D., Zanker, M., Felfernig, A., Friedrich, G.: Recommender Systems: An Introduction. Cambridge University Press, New York (2010)

Kaplan, A.M., Haenlein, M.: Users of the world, unite! The challenges and opportunities of Social Media. Bus. Horiz. **53**(1), 59–68 (2010)

Koren, Y., Bell, R., Volinsky, C.: Matrix factorization techniques for recommender systems. Computer **42**(8), 30–37 (2009)

Lee, W.-P., Liu, C.-H., Lu, C.-C.: Intelligent agent-based systems for personalized recommendations in Internet commerce. Expert Syst. Appl. **22**(4), 275–284 (2002)

Lekakos, G., Caravelas, P.: A hybrid approach for movie recommendation. Multimedia Tools Appl. **36**(1–2), 55–70 (2008)

Linden, G., Smith, B., York, J.: Amazon.com recommendations: Item-to-item collaborative filtering. IEEE Internet Comput. **7**(1), 76–80 (2003)

Liu, N.N., He, L., Zhao, M.: Social temporal collaborative ranking for context aware movie recommendation. ACM Trans. Intell. Syst. Technol. (TIST) **4**(1), 15 (2013)

Lu, Y., Wang, F., Maciejewski, R.: Business intelligence from social media: A study from the vast box office challenge. IEEE Comput. Graph. Appl. **34**(5), 58–69 (2014)

McSherry, F., Mironov, I.: Differentially private recommender systems: building privacy into the net. In: Proceedings of the 15th ACM SIGKDD International Conference on Knowledge Discovery and Data Mining, pp. 627–636. ACM, New York (2009)

Said, A., Berkovsky, S., De Luca, E.W., Hermanns, J.: Challenge on context-aware movie recommendation: CAMRa2011. In: Proceedings of the Fifth ACM Conference on Recommender Systems, pp. 385–386. ACM, New York (2011)

Sarwar, B., Karypis, G., Konstan, J., Riedl, J.: Analysis of recommendation algorithms for e-commerce. In: Proceedings of the 2nd ACM Conference on Electronic Commerce. ACM (2000)

Sarwar, B., Karypis, G., Konstan, J., Riedl, J.: Item-based collaborative filtering recommendation algorithms. In: Proceedings of the 10th International Conference on World Wide Web, pp. 285–295. ACM, New York (2001)

Schafer, J.B., Konstan, J.A., Riedl, J.: E-commerce recommendation applications. Applications of Data Mining to Electronic Commerce, pp. 115–153. Springer (2001)

Szomszor, M., Cattuto, C., Alani, H., O'Hara, K., Baldassarri, A., Loreto, V., Servedio, V.D.: Folksonomies, the semantic web, and movie recommendation. In: 4th European Semantic Web Conference, Bridging the Gap between Semantic Web and Web 2.0, Innsbruck, Austria, pp. 1–14 (2007)

Zimmerman, J., Kauapati, K., Buczak, A.L., Schaffer, D., Gutta, S., Martino, J.: TV personalization system. In: Personalized Digital Television, pp. 27–51. Springer (2004)

Predicting the Scale of Trending Topic Diffusion Among Online Communities

Dohyeong Kim[1], Soyeon Caren Han[2], Sungyoung Lee[1],
and Byeong Ho Kang[2(✉)]

[1] Department of Computer Engineering,
Kyung Hee University, Youngin, Giheung-gu, Korea
{dhkim,sylee}@oslab.khu.ac.kr
[2] School of Engineering and ICT, University of Tasmania,
Sandy Bay, Tasmania 7005, Australia
{Soyeon.Han,Byeong.Kang}@utas.edu.au

Abstract. Online trending topics represent the most popular topics among users in certain online community, such as a country community. Trending topics in one community are different from others since the users in the community may discuss different topics from other communities. Surprisingly, almost 90 % of trending topics are diffused among multiple online communities, so it shows peoples interests in a certain community can be shared to others in another community. The aim of this research is to predict the scale of trending topic diffusion among different online communities. The scale of diffusion represents the number of online communities that a trending topic diffuses. We proposed a diffusion scale prediction model for trending topics with the following four features, including community innovation feature, context feature, topic feature, and rank feature. We examined the proposed model with four different machine learning in predicting the scale of diffusion in Twitter Trending Topics among 8 English-speaking countries. Our model achieved the highest prediction accuracy (80.80 %) with C4.5 decision tree.

Keywords: Twitter · Twitter trending topics · Information diffusion · Trending topic diffusion · Diffusion prediction

1   Introduction

By using different types of web-based services, such as search engines, social media, and Internet news aggregation sites, internet users can share and search information throughout the world. These web services, including Google, Yahoo, and Twitter, analyse their social data and provide a Trending Topics service, which displays the most popular terms that are discussed and searched within their community. One of these web services, Twitter, monitors their social data, detects the terms (including phrases and hash-tags) currently most often mentioned by their users, and publishes these on their site. Abdur Chowdhury[1],

---

[1] Twitter, Inc. 2009 https://blog.twitter.com/2009/top-twitter-trends-2009.

© Springer International Publishing Switzerland 2016
H. Ohwada and K. Yoshida (Eds.): PKAW 2016, LNAI 9806, pp. 153–165, 2016.
DOI: 10.1007/978-3-319-42706-5_12

a chief scientist at Twitter Research Team, defined the Twitter Trending Topics as the objects showing us that people everywhere can be united in concern around important events. Trending topics are estimated to reflect the real-world issues from the people's point of view. Kwak et al. [13] demonstrated that over 85 % of trending topics in Twitter are related to breaking news headlines, and the related tweets of each trending topic provide more detailed information of news and users' opinions. Hence, being able to know which topics people are currently most interested in on Twitter, and their point of view, may lead to opportunities for analysing the market share in almost every industry or research fields, including marketing, politics, and economics.

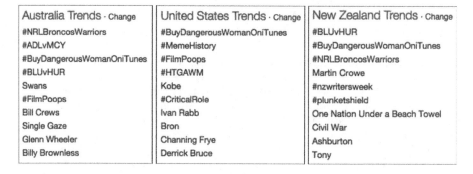

Fig. 1. Twitter trending topics for different countries

The 'Twitter Trending Topics' service includes the top 10 trending topics for different locations. This list enables recognition of the degree of current popularity of that topic in a specific geographic location, from individual cities, to countries, and worldwide[2]. Figure 1 shows the top 10 trending topics for Australia, United States, and New Zealand at 4pm, 11th March 2016. The trending topics in one country's community are different from other countries' communities. This is because peoples' interests differ across countries. The Trending Topics list for each country includes their local events and worldwide events.

The interesting phenomenon is that many trending topics tend to diffuse among multiple countries. As can be seen in Fig. 2, we found that over 90 % of trending topics for each country appeared in different countries trending topics list. 92.27 % of the U.Ks trending topics also appeared in at least two different countries' trending topic lists, while only 7.73 % of them appeared solely in the U.Ks trending topics list. This shows that the majority of trending topics appeared and diffused in multiple countries' trending topics. For example, on July 17th 2014, when a missile downed the Malaysia Airlines plane over Ukraine, it was breaking news around the world. During this time, the topic '#MalaysiaAirlines'

---

[2] Twitter, Inc. 2014 https://support.twitter.com/articles/101125-faqs-about-trends-on-twitter.

initially appeared on the Malaysian trending topics list. After that, Twitter users in the Philippines, Singapore, Australia, and New Zealand started talking about the trending topic. Finally, 3 hours after starting in the Malaysian trending topics list, all 8 English speaking countries were discussing the topic '#MalaysiaAirlines' because it is world breaking news. If the trending topic is just about local events, it normally diffuses to less than two countries.

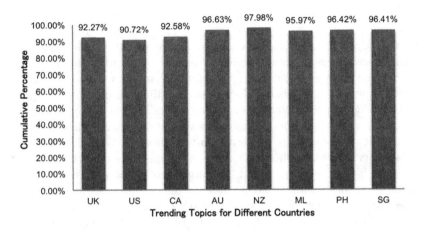

Fig. 2. Percentage of trending topics that diffused Percentage

The number of countries that trending topics diffuse to can be the barometer of people's interests in trending topics. We defined the number of diffused countries as the scale of diffusion for trending topics. Predicting the scale of diffusion for trending topics can be helpful to identify the influence of the topic in the near future.

However, the 'Trending Topics' list displays only limited information, including the trending topic term, its rank, location, and updated date and time. We used only this available information to build the diffusion scale prediction model for trending topics among different online communities. Therefore, the research aim of our study is to answer the following question: "How can we predict the scale of diffusion for trending topics with the available, limited information?" In this paper, we focused on finding the important features that can be related to the scale of trending topics, and building the successful prediction model using machine learning techniques.

The paper is structured as follows: Sect. 2 presents the related work, followed in Sects. 3 and 4 by the methodology for collecting trending topics and building the Diffusion Scale Prediction Model for trending topics. In Sect. 5, we describe the evaluations conducted, and discuss the results. Finally, we conclude this paper in Sect. 6.

2   Related Works

In order to build the information diffusion model for the online trending topics, we conducted the literature review in several related research areas, including online trending topics and information diffusion. This section covers how the trending topics are extracted from the online environment, how trending topics are used in the real world, and the different types of information diffusion model proposed in other research areas.

Trending Topic Extraction in Online Environment. Trending topics are the topic or issue that a great number of people is talking about in a certain period of time. In the early stage, the trending topics tend to be extracted from newspapers or paper-based documents. In 1992, Andersen et al. [2] proposed JASPER, which extracts the trending topics from newspapers and financial reports in order to support decision-making in a finance environment. Since 2000, Internet users can share and search information by using different types of web-based services and applications, such as Google, Yahoo, Twitter, or Weibo [9]. Those web-based services found that user activity data represents the peoples interests in the certain period, so they started detecting the peoples interests and presenting the most discussed and searched topics, Trending Topics, from their users data. For example, Twitter, one of the most popular social media websites, analyses postings from more than 41 million users and provides the real-time top 10 trending topics of different regions from cities to worldwide [13]. Weibo, a popular Chinese social media website, also presents the top 10 popular topics with the number of searches [4]. In addition to social media, search engines such as Google and Baidu, analyse their users search activities, and share the list of daily popular search keywords that represents peoples interests on a day [8]. Since people tend to make a decision relying on the mainstream interests and behaviors, the Trending Topic services received a lot of attention from researches and industries.

Research Using Topics in Online Environment. Many researchers applied various summarisation and extraction approaches aimed at revealing the exact meaning of trending topics. Sharifi [17] applied a phrase reinforcement algorithm to summarise related tweets of Twitter Trending Topics. Then, the author conducted evaluation for comparing hybrid TFIDF and phrase reinforcement in use of Trending topics summarising. Inouye [11] also conducted an experiment to compare twitter summarisation algorithms. They found that simple frequency-based techniques produce the best performance in tweets summarisation. Han and Chung [6] applied simple Term Frequency approach for extracting the representative keywords to disambiguate the approach. They also proved that the most successful approach to reveal the exact meaning of trending topics is simple Term Frequency, which is evaluated by 20 postgraduate students [7]. Some researchers examined classifying trending topics. Lee et al. [14] classifies trending topics into 18 general categories by labeling and applying machine-learning techniques. Zubiaga et al. [18] aimed to classify trending topics by applying several proposed features and used SVM to check the accuracy. Han et al. [10] proposed a temporal modeling framework for predicting rank change of trending topics using historical rank data. The proposed model achieved extremely high prediction accuracy (94.01 %) with a C4.5 decision tree.

Information Diffusion Modelling. We conducted reviews of several Information Diffusion Models used in predicting product diffusion, blog posting diffusion, and social media posting diffusion. The early diffusion models are derived from the Bass diffusion model [12], which calculates the diffusion using the relationship between the current adopters and potential adopters with a new product interaction. Rogers published diffusion of innovations to against a typographical error of Bass paper, which did not cover how, why, and at what rate new ideas and technology spread of a new idea. According to Roger [16], diffusion is a process by which an innovation is communicated through certain channels over time among members of a social system, and the member can be classified into five categories, including innovators, early adopters, early majority, late majority, and laggard.

3   Trending Topics Monitoring

In this paper, we focused on predicting the scale of trending topics diffusion across multiple countries. For our studies, we collected Twitter Trending Topics from 8 different countries communities, including U.S., U.K., Australia, New Zealand, Canada, Malaysia, Philippine, and Singapore. Twitter provides an API (Application Programming Interface) that allows developers or researchers to crawl and collect the data easily. Through this API service, we collected twitter trending topics for 3 years (until 31th December, 2015)

3.1   Trending Topics

Twitter monitors all users data and detects the popular trending topics that most people are currently discussing about. The detected popular trending topics are displayed on the service 'Twitter Trending Topics'. This trending topic service is located on the sidebar of Twitter interface by default so it is very easy for users to check the current trending topics and discuss about it. It provides top 10 trending topics in real time. Hence, we have collected those top 10 trending topics per hour using Twitter API. In total, we have collected 705354 unique trending topics in 3 years.

3.2   Related Tweets

Trending topics in Twitter consist of short phrases, words, or hash-tags. Twitter never provides any detailed explanation of trending topics so it is very difficult to identify the meaning of trending topics until you have a look related tweets of those topics. For example, when a missile destroys Malaysian Airlines, the trending topics were 'Malaysia Airlines', 'Malaysian', etc. It is almost impossible to realise what happened to the Malaysia Airlines by only checking the trending topics. In order to reveal the exact meaning of each trending topic, we need to collect not only the trending topics, but also the appropriate related tweets of a specific trending topic, and the related tweets should not contain contents that are irrelevant. If the trending topic is Malaysia Airline which is about a missile attack that happened on July 18th, we should not collect the related tweets about a missing Malaysia Airliner that occurred on March 8th. It is extremely important to distinguish the tweets that are related to specific trending topics. Twitter API provides the tweet/search crawling service that allows users to collect the tweets by using the search query. The concept of the tweet/search service is the same as a search engine. Users can search the tweets that contain the search keyword. The search results contain detailed information about each tweet, including content, username, location, created date-time, etc. We used the created date-time to extract the appropriate tweets for the trending topics. As we collect the top 10 trending topics on an hourly basis, we search and collect the related tweets that users upload in the last one hour. For example, when Malaysia Airline is on the trending topics list at 8pm, we search and collect the related tweets that users upload in the previous one hour, 7pm to 8pm. This collection approach prevents irrelevant tweets.

4   Trending Topics Diffusion Prediction Model

The goal of this research is to predict the scale of trending topics diffusion across multiple countries. We propose a model that predicts the number of countries the trending diffuses by using four different trending topic features and machine learning technique. The proposed model can be described using the following equation:

$$Scale(T_x) = ML(DP(T_x)) \tag{1}$$

$$DP(T_x) = [CI(T_X), CT(T_X), TP(T_X), RK(T_X)] \tag{2}$$

In order to predict the scale $Scale$ of the trending topic T_x diffusion, we extracted four different features of the topic T_x for the diffusion prediction DP model. As can be seen in Eq. 2, the four features of the specific trending topic T_x include community innovation level $CI(T_X)$, context feature $CT(T_X)$, topic feature $TP(T_X)$, and rank feature $RK(T_X)$. Then, machine learning techniques ML are applied for learning our model. $Scale$ represents the number of countries the trending topic T_x diffuses to. The outcome/result will be the number, from 1 to 8. For example, if the prediction is 1, it identifies that there will be no diffusion to other countries.

$$Scale(T_x) = \begin{cases} 1, \text{will be no diffusion} \\ ... \\ 8, \text{will be diffused to all countries} \end{cases} \tag{3}$$

4.1   Community Innovation Level Feature

Community innovation level feature of the trending topic determines an innovation level of online community that the topic initiates. We modified the concept of innovation of diffusion approach proposed by Roger [16]. This feature classifies communities based on the level of the community adopting the trending topic. There are four types of levels: (1) Innovator: Communities that start diffusing the trending topics, (2) Early Adopter: Communities that adopt the diffused trending topics in the early stage, (3) Late Adopter: Communities that adopt the diffused trending topics after the average participant, and (4) Laggards: Communities that are the last to adopt the diffused trending topic.

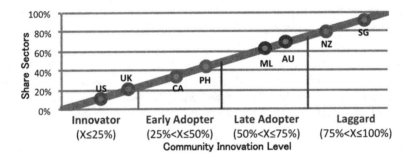

Fig. 3. Community innovation level of 8 different English speaking countries (Color figure online)

The way we classify the community innovation feature can be seen from Fig. 3 the graph shows the innovation level of each country community. For example, U.K. and U.S. are the Innovator, which tend to start diffusing the trending topics to other countries' communities, and Canada (CA) and Philippines (PH) are the Early Adopter that usually adopt the trending topics in the early stage. The Y-axis shows the percentage of share sectors (market share the percentage of people who know the specific trending topics), and the X-axis represents the average time taken to adopt the specific trending topics. We classified the innovation levels based on the average percentage of time that a country spent on adopting the trending topics. For example, on average, when only 15 % of 8 English-speaking countries communities adopts the certain trending topic, the topic is already extremely popular among majority of people in U.S. online community and appears in U.S. trending topic list.

4.2   Context Feature

This feature enables identification of the type of trending topics by using context patterns. We used three categories, including breaking news, meme, and commemorative day, on the context patterns. Table 1 shows the example pattern for classifying the type of trending topics. Based on this context pattern and the type, we created and used over 20 rules to classify the trending topics using context patterns. For example, if the trending topic contains Person Pronoun AND Noun, then the trending topic is classified into Meme category.

Table 1. Context Feature Classification Pattern Example

Patterns	Example	Category
Be + Noun	Nothing was the same	Meme
Verb + Noun	#HowToAskSomeoneOnADate	Meme
Person pronoun + Noun	#ILoveEXO	Meme
Noun + Commemorative days	Christmas	Commemorative
Noun + No Commemorative days	Galaxy	News

Table 2 shows that over 85 % of trending topics collected in three years are talking about the news, which is matched with the results from Kwak et al. [1]. They mentioned that around 80 % of trending topics are related to the title of breaking news.

Table 2. Percentage of trending topics based on the context pattern

Context categories	Percentage
Commemoratives	4 %
Meme	9 %
News	87 %

4.3   Topic Feature

This feature represents the semantic topic of the Trending Topics. As trending topics are mostly about new and real-time events, the traditional topic classification ontology cannot be applied. This is because traditional ontology usually does not contain new terms, and it normally classifies the topic into the semantically related category, rather than the category related to the real-time situation about the topic. Assume the topic 'Samsung' is on the 'Trending Topic' list now. In this case, traditional topic classification ontology will classify the topic into 'Technology'. However, if the trending topic 'Samsung' is about the news, Samsung sponsoring British football team Chelsea, the topic should be classified into 'Sports' category.

Therefore, we classified the trending topics using the New York Times (NY times) classification service. The service provides nine (9) topic categories as follows: Sports, Entertainment, Politics, Business, World issue, Technology, Fashion, Obituaries, Health. The way we identified the category of a trending topic is as follows: first of all, we search the trending topic with the NY times topic classification service. We set the published time as the day that trending topic first emerged. Then we can locate any related articles that were published with that term, on that day. Finally, the trending topics related categories are supplied by the NY Times classification service. As can be seen in the Table 3, we found that 80 % of trending topics are classified in the following three categories, including entertainment, sports and politics. It represents that most people are interested in the issues/events of entertainment, sports, or politics.

Table 3. Topic distribution in U.S. Trending Topics

No	Topic	Percentage
1	Entertainment	42 %
2	Sports	28 %
3	Politics	10 %
4	Fashion	6 %
5	World issue	5 %
6	Obituaries	4 %
7	Health	3 %
8	Technology	2 %

4.4 Rank Feature

Each trending topic has a popularity ranking (Rank 1 to 10). The ranking of trending topics is changing in real-time. For this feature, we used the ranking value of a trending topic when it was initiated/started from a certain country. However, there are some trending topics, which are starting in the multiple countries' communities. In this case, we checked the community innovation level of starting countries. Once a trending topic comes in, first check whether it started in a single country. When it starts in a single country, then we use the original rank of this trending topic. When it initiates in multiple countries, then check whether it starts from the country that is innovator level. If it starts from innovator, then check whether there are more than one innovator for this trending topic. If more than one, then choose the highest rank of this trending topic as the starting rank. If there is only one innovator, then choose rank from innovator as starting rank of this trending topic. If there is no innovator, then choose the highest rank of this trending topic as the starting rank.

5   Evaluation

5.1   Evaluation Data

As mentioned in Sect. 3, we used Twitter API and collected trending topic terms, related tweets and ranking patterns for those topics for three years (from 1st January, 2013 to 31st December, 2015) in 8 different English speaking countries. We crawled the top 10 trending topics for each country every hour. The API returns the trending topic term, the rank, the location and time of the API request. The trending topics list displays only the topics terms with no detailed information so we searched the related tweets of each trending topic and used published date-time to extract the appropriate related tweets.

In order to achieve the Eq. 1 in Sect. 4, we used the training data contains trending topics with extracted four different features (including Community Innovation level, Context, Topic, and Rank) as attributes and the number of diffused countries as a class/outcome.

5.2   Machine Learning Techniques

For building the prediction model using our training data, we applied machine learning techniques, which are initially introduced for predictions. We selected four machine learning techniques: Naive Bayes(NB) [5], Neural Networks(NN) [3], Support Vector Machines(SVM) [1], and Decision Trees(DT) [15]. The philosophies behind these techniques are very different, but each has been shown to be effective in several time-series prediction studies.

5.3   Evaluation Result

We used 10-fold cross validation on three years of training data, which is described in Sect. 5.1. The main objective of this experiment is to examine the scale prediction model of trending topics diffusion in four machine learning techniques. Experiment demonstrated that the proposed diffusion prediction model reasonably predicts the scale of diffusion of trending topics. Table 4 presents the scale of diffusion prediction accuracy with four machine learning techniques. The combination of context and topic features produce the lowest accuracy with the four techniques. In contrast, using only community level or ranking feature, provides a better accuracy with both being over 0.5. The accuracy results increased from 0.04 up to 0.07 by using community level feature and ranking feature together. The greatest accuracy was achieved by using all four features of evaluation in the four machine learning techniques, this being over 0.7, which is much higher than using a single feature, or two features, to predict the scale of trending topic diffusion.

By using the combination of context and topic features, the scale of diffusion prediction accuracy just reached 0.28, which is not enough for predicting the scale of trending topics. This result can be explained in two reasons: First, more than 80 % of trending topics are categorised into the 'News' type by using

Table 4. The prediction accuracy with four machine learning techniques

	Context + Topic	Rank	Community Level	Rank + Community Level	All
NB	0.212	0.582	0.545	0.632	0.722
NN	0.252	0.592	0.559	0.636	0.738
SVM	0.258	0.593	0.558	0.642	0.727
C4.5	0.281	0.609	0.658	0.736	0.808

given context pattern rules so that the context feature is not suitable to classify or predict the scale of trending topics. Secondly, topic feature has similar circumstances, which are usually classified trending topics into the following three groups, including sports, entertainments, and politics. Therefore, it is not suitable to use only context or topic feature as a parameter for predicting the scale of trending topic diffusion.

The prediction accuracy obtained by using only ranking or community level feature in NN and SVM are almost the same. However, the prediction result with the combination of ranking and community level feature in NN is lower than those in SVM, and the accuracy result of using all four factors in NN is higher than in SVM. SVM can solve the problem of structure selection in NN, which can improve the accuracy result of using ranking and community level factors. Comparing to NN, the accuracy of using context factor is better in SVM, while the performance outcome of using all four factors is lower in SVM. That means patterns for adding context and topic factors better fit the NN algorithm. Another interesting finding is that C4.5 decision tree have better accuracy than the other three machine learning techniques. Since C4.5 decision tree uses a tree to conduct the prediction/classification, the classifying results may better fit with the proposed diffusion scale prediction model. When we use context and topic factor with C4.5, the accuracy result is lower than the others, just reaching 0.281. However, using only community level factor or ranking factor, the accuracy result

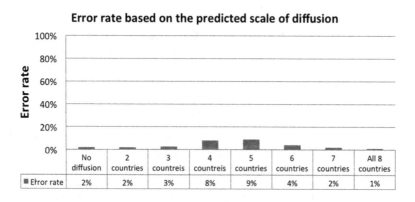

Fig. 4. Error rate based on the predicted scale of diffusion

reaches over 0.6. When combining ranking factor and community level factor, the accuracy result increased at least 0.07 compared to using a single factor. The highest accuracy result almost reached 0.8 by using all four factors together.

Figure 4 represents the error rate based on the predicted scale of trending topics diffusion. The error rate shows that the prediction model is successfully predicting when the trending topic has less diffusion or diffuses to all communities. Based on our analysis, 50 % of trending topics tend to diffuse 2 or 8 countries. Therefore, for the future work, this would require an analysis of trending topics in order to identify the important features in distinguishing the medium level of popular trending topic diffusion patterns.

6   Conclusion

In this paper, we proposed a diffusion scale prediction model for trending topics among different online communities. We developed the model with four important features that can be used in predicting the number of countries that trending topics diffuses to. The four features include community innovation level, context, topic, and rank features. The proposed features are learned by four different machine-learning techniques: Nave Bayes, Neural Network, Support Vector Machine, and C4.5 decision tree. Based on the results, the prediction model learned by C4.5 decision tree achieved the highest prediction accuracy (80.8 %) in scale prediction. Compared to traditional social data applied diffusion prediction model, our proposed prediction model works successfully. For the future work, we will focus on predicting range (depth) and speed of trending topic diffusion so it can forecast three dimensions of trending topic diffusion patterns. We hope the paper is the step forwards improving the performance for any researches using trending topics as a data.

References

1. Abe, S.: Support Vector Machines for Pattern Classification. Springer, London (2010)
2. Andersen, P.M., Hayes, P.J., Huettner, A.K., Schmandt, L.M., Nirenburg, I.B., Weinstein, S.P.: Automatic extraction of facts from press releases to generate news stories. In: Proceedings of the Third Conference on Applied Natural Language Processing, pp. 170–177. Association for Computational Linguistics, March 1992
3. Araujo, P., Astray, G., Ferrerio-Lage, J.A., Mejuto, J.C., Rodriguez-Suarez, J.A., Soto, B.: Multilayer perceptron neural network for flow prediction. J. Environ. Monit. 13(1), 35–41 (2011)
4. Chen, L., Zhang, C., Wilson, C.: Tweeting under pressure: analyzing trending topics and evolving word choice on sina weibo. In: Proceedings of the First ACM Conference on Online Social Networks, pp. 89–100. ACM, October 2013
5. Efron, B.: Bayes' theorem in the 21st century. Science 340(6137), 1177–1178 (2013)
6. Han, S.C., Chung, H.: Social issue gives you an opportunity: discovering the personalised relevance of social issues. In: Richards, D., Kang, B.H. (eds.) PKAW 2012. LNCS, vol. 7457, pp. 272–284. Springer, Heidelberg (2012)

7. Han, S.C., Chung, H., Kim, D.H., Lee, S., Kang, B.H.: Twitter trending topics meaning disambiguation. In: Kim, Y.S., Kang, B.H., Richards, D. (eds.) PKAW 2014. LNCS, vol. 8863, pp. 126–137. Springer, Heidelberg (2014)
8. Han, S.C., Liang, Y., Chung, H., Kim, H., Kang, B.H.: Chinese trending search terms popularity rank prediction. Inf. Technol. Manag. 1–7 (2015)
9. Han, S.C., Kang, B.H.: Identifying the relevance of social issues to a target. In: 2012 IEEE 19th International Conference on Web Services (ICWS), pp. 666–667. IEEE, June 2012
10. Han, S.C., Chung, H., Kang, B.H.: Trending topics rank prediction. In: Wang, J., Cellary, W., Wang, D., Wang, H., Chen, S.-C., Li, T., Zhang, Y. (eds.) Web Information Systems Engineering – WISE 2015. LNCS, vol. 9419, pp. 316–323. Springer, Heidelberg (2015)
11. Inouye, D., Kalita, J.K.: Comparing twitter summarization algorithms for multiple post summaries. In: 2011 IEEE Third International Conference on Privacy, Security, Risk and Trust (PASSAT), and 2011 IEEE Third International Conference on Social Computing (SocialCom), IEEE (2011)
12. Jeyaraj, A., Sabherwal, R.: The bass model of diffusion: recommendations for use in information systems research and practice. JITTA: J. Inf. Technol. Theor. Appl. 15(1), 5 (2014)
13. Kwak, H., Lee, C., Park, H., Moon, S.: What is Twitter, a social network or a news media? In: Proceedings of the 19th International Conference on World Wide Web, pp. 591–600. ACM, April 2010
14. Lee, K., Palsetia, D., Narayanan, R., Patwary, M.M.A., Agrawal, A., Choudhary, A.: Twitter trending topic classification. In: 2011 IEEE 11th International Conference on Data Mining Workshops (ICDMW), pp. 251–258. IEEE, December 2011
15. Loh, W.Y.: Classification and regression trees. Wiley Interdisc. Rev. Data Min. Knowl. Disc. 1(1), 14–23 (2011)
16. Rogers, E.M.: Diffusion of Innovations. Simon and Schuster, Chicago (2010)
17. Sharifi, B., Hutton, M.A., Kalita, J.: Summarizing microblogs automatically. In: Human Language Technologies: The 2010 Annual Conference of the North American Chapter of the Association for Computational Linguistics, pp. 685–688. Association for Computational Linguistics, June 2010
18. Zubiaga, A., Spina, D., Fresno, V., Martinez, R.: Classifying trending topics: a typology of conversation triggers on twitter. In: Proceedings of the 20th ACM International Conference on Information and Knowledge Management, pp. 2461–2464. ACM, October 2011

Finding Reliable Source for Event Detection Using Evolutionary Method

Raushan Ara Dilruba[1,2] and Mahmuda Naznin[1(✉)]

[1] Department of CSE, Bangladesh University of Engineering and Technology,
Dhaka, Bangladesh
mahmudanaznin@cse.buet.ac.bd
[2] Department of CS, University of Calgary, Calgary, Canada
rdilruba@ucalgary.ca

Abstract. Participatory sensing is a phenomenon where participants use mobile phones or social media and feed data to detect an event. Since, data gathering is open to many participants, one of the major challenges of this type of networks is to identify truthfulness of the reported observations. Finding the reliable sources is a challenging task since the node or participant's reliability is unknown or even the probability of the reported event to be true is also unknown. In our paper, we study this challenge and observe that applying evolutionary method, we can identify reliable source nodes. We call our approach Population Based Reliability Estimation. We validate our claim by experimental results. We also compare our method with another widely used method. From experiments we find that our approach is more efficient.

Keywords: Evoloutinray approach · Reliability · Participatory sensing · Genetic Algorithm

1 Introduction

Participatory Sensing is a process of data collection and interpretation of an event by feeding interactive data via web or social media [28,31]. A *Participatory Sensor Network* consists of nodes or participants to collect data for a common project goal within its framework [1,3]. The nodes or participants use their personal mobile phones to sense various activities of their surrounding environment and submit sensed data through mobile network or social networking sites [3,24,25].

However, finding reliable sources in a participatory sensor network is very challenging task due to the big and continuous volume of sensing and communication data generated by the participant nodes and the availability of ubiquitous, real-time data sharing opportunities among nodes [2,12,16,17,20]. One conventional way to collect the reliable data is conducting self-reported surveys. However, conducting survey is a time consuming procedure. Popular data collection can be achieved by using mobile devices like smart phones, wearable sensing

© Springer International Publishing Switzerland 2016
H. Ohwada and K. Yoshida (Eds.): PKAW 2016, LNAI 9806, pp. 166–180, 2016.
DOI: 10.1007/978-3-319-42706-5_13

devices or through social networks [4, 10, 23, 26]. In a participatory network, the users are considered as participatory sensors, and an event can be reported or detected by the users [21, 22]. The major challenge in this participatory sensing is to ascertain the truthfulness of the data and the sources. The reliability of the sources is questionable because the data collection is open to a very large population [36]. The *reliability* of the participants (or sources) denotes the probability that the participant reports correct observations. Reliability may be impaired because of the lack of human attention to the task, or because of the bad intention to deceive. Without knowing the reliability of sources, it is difficult to measure whether the reported observations or events are true or not [32]. Openness in data collection also leads to numerous questions about the quality, credibility, integrity, and trustworthiness of the collected information [5, 9, 14, 15]. It is very challenging to find whether the end user is correct, truthful and trustworthy. If the nodes are reliable, the credibility of the total system increases. Therefore, it is very important to find the reliable sensing sources to detect the events.

In this paper, we address the challenges of finding the node reliability in a participatory sensing system. Our paper is organized as follows. Section 2 provides the background study. Section 3 provides the problem domain in details. Section 4 provides the experimental results. Finally, Sect. 5 gives the conclusion of this research work.

2   Related Work

In this section, we discuss some research work on the reliability estimation of the nodes in a participatory sensing network. For the case of specific kinds of data such as location data, a variety of methods are used in order to verify the truthfulness of the location of a mobile device [19]. The key idea is that time-stamped location certificates signed by wireless infrastructure are issued to co-located mobile devices. A user can collect certificates and later provide those to a remote party as a verifiable proof of his or her location at a specific time. However, the major drawback of this approach is that the applicability of these infrastructure based approaches for mobile sensing is limited as cooperating infrastructure may not be present in remote or hostile environments.

In the context of participatory sensing, where raw sensor data is collected and transmitted, a basic approach for ensuring the integrity of the content has been proposed in [14], which guards whether the data produced by a sensor has been maliciously altered by the users. Trusted Platform Module (TPM) hardware [14], can be leveraged to provide this assurance. However, this method is expensive and not practical since each user must have predefined hardware framework.

The problem of trustworthiness has been studied for resolving multiple, conflicting information on the web in [36]. The earliest work in this regard are proposed in [7, 8]. A number of recent methods in [3, 18, 32, 33] also address this issue, in which a consistency model is constructed in order to measure the trust in user responses in a participatory sensing environment. The key idea of this system is that untrustworthy responses from users are more likely to be different from one another, whereas truthful methods are more likely to be consistent with one another. This broad principle is used in order to model the likelihood of participant reliability in social sensing with the use of a Bayesian approach [32]. A system called Apollo [18] has been proposed in this context in order to find the truth from noisy social data streams. However, these types of methods are a bit time consuming and do not ensure source reliability. In case of collaborative attack this method may fail.

In [18], authors present a fuzzy approach where this system is able to quantify uncertain and imprecise information, such as trust, which is normally expressed by linguistic terms rather than numerical values. However, the linguistic term can create vague results. In [28,34], authors present a streaming approach to solve the truth estimation problem in crowd sourcing applications. They consider a category of crowd sourcing applications, the truth estimation problem. This is basically reliability finding problem. In fact fact-finding algorithms are used to solve this problem by iteratively assessing the credibility of sources and their claims in the absence of reputation scores. However, such methods operate on the entire dataset of reported observations in a batch fashion, which makes them less suited to applications where new observations arrive continuously. The problem is modelled as an Expectation Maximization (EM) problem to determine the odds of correctness of different observations [3]. Problem and accessing on-line information from various data sources are mentioned in [37]. However, all of these methods suffer from the collecting the ground truth data for unavoidable circumstances. Therefore, finding credible nodes of the detected event is still very challenging research problem.

3   Problem Domain

In this section, we define the system model, problem formulation and give the details of our methodology. At first we discuss some preliminaries relevant to our research problem. Let us consider a participatory sensing model where a group of M participants, $S_1 ... S_M$, make individual observations about a set of N events $C_1 ... C_N$. The probability that participant S_i reports a true event when the event is actually true is a_i and the probability that participant S_i reports a true event when the event is actually false is b_i. θ is the set of a_i and b_i, $\theta(a_i, b_i)$.

To handle this challenge, we apply Genetic Algorithm or Population Based Method where we can keep around a sample of candidate solutions rather than a single candidate solution to find the solution quickly [11]. We know, GA generate solutions to optimize the problems inspired by natural evolution, such as *inheritance, mutation, selection,* and *crossover* [11]. We call our method *Population Based Reliability Estimation (PBRE)* which uses a set of reliability for the population instead of single reliability. In our approach, we call this set of reliability as P and we use Genetic Algorithm to estimate the best possible reliable participants. We call the set of θ as P which is a set of reliability. z_j is the probability that the event or claim C_j is indeed authentic.

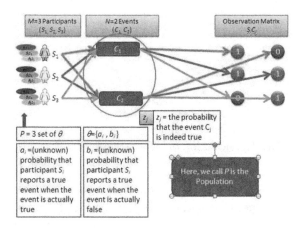

Fig. 1. Population based system model.

3.1 Population Based Method

In this section, we provide an outline of this method as follows.

Step 1: We initialize and build population in the following ways:

1. We initialize M, N
2. We take input SC matrix or $Source-Claim$ matrix. Each entry of the matrix is either 0 or 1. Here, when the participant S_i reports an event C_j as false $S_iC_j = 0$, and when S_i reports an event C_j as true $S_iC_j = 1$. We assume, each observation and source of the matrix is independent of each other.
3. We initialize d, overall bias on event to be true (value may range from 0 to 1).
4. Finally, $P=$ The set of θ āny value between 0 to 1.

Step 2: We calculate z_j as follows.

$p(z_j|X_j, \theta)$ is the conditional probability to be true, given the SC matrix X_j related to the jth and the current estimate of θ.

Step 3: We compute fitness. *Computing Fitness* is described as follows. Then, we assess fitness of P, the set of reliability. We compare P with the best reliability. The *target reliability* or $target\_a_i$ is computed as follows.

$$target\_a_i = \sum_{i=1}^{M}(\sum_{j=1}^{N} = \frac{S_i \times C_j}{N})$$

For example, in an ideal case, when the probability of all events to be true, $z_j = 1$. Let us consider, there are 2 events and 3 participants which is illustrated in Fig. 2. Participant S_1 reports event C_1 as true, and reports the event S_2 as false. Therefore, $target\_a_1 = \frac{S_1C_1 + S_1C_2}{2} = \frac{1+0}{2} = 0.5$.

Now, the objective is to select the best fit or the fittest a_i from P that helps to converge z_j. We take the fittest value from the initial set of values of a_i using the fitness function. We call this fittest value as *fit reliability* or fit_a_i.

Now, we define two types of fitness functions *Fit_Parent* and *Replace_Parent*.

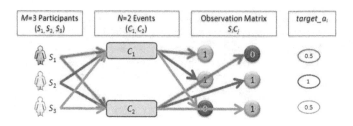

Fig. 2. Calculating the most reliable $target\_a_i$

Type 1 : *Fit_Parent-* *Fit_Parent* selects fit_a_i from the set of a_i of S_i. Here, fit_a_i is the closest value to $target\_a_i$. We describe the computation in Fig. 3.

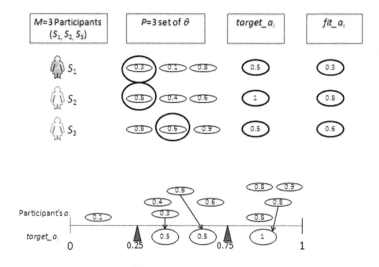

Fig. 3. Fit_Parent computation

Fit_Parent selects fit_a_i from the set of a_i of S_i. Here, fit_a_i is the closest value to $target\_a_i$. We describe the computation in Fig. 3. For example, we initialize three sets of a_1 for participant S_1 e.i. 0.3, 0.1 and 0.8. Figure 3 is an illustrative example of Fit_Parent computation. We see that the $target\_a_1$ is 0.5. Therefore, the closest a_1 e.i. fit_a_1 is 0.3. Similarly, we calculate fitness for participant S_2 and S_3 which are $a_2 = 0.8$ and $a_3 = 0.6$ respectively.

Type 2 : *Replace_Parent-* Here, instead of selecting one fit_a_i from every participant S_i's P, we select the full set of a_i which is the closest to set of $target\_a_i$. Now, we give an illustrative example of Replace_Parent in Fig. 4.

For example, we initialize three sets of a_i for each participant S_1 e.i. $(a_{11}, a_{12}, a_{13}) = (0.3, 0.1, 0.8)$, for S_2 it is $(a_{21}, a_{22}, a_{23}) = (0.8, 0.4, 0.5)$ and for S_3 it is $(a_{31}, a_{32}, a_{33}) = (0.8, 0.5, 0.9)$. Now, we make another set taking the first a_i

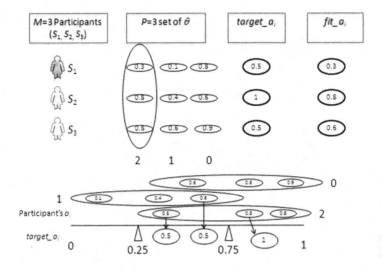

Fig. 4. Replace_Parent computation

from each S_i e.i. $(a_{11}, a_{21}, a_{31}) = (0.3, 0.8, 0.8)$ and similarly $(a_{12}, a_{22}, a_{32}) = (0.1,$ $0.4, 0.6)$ and $(a_{13}, a_{23}, a_{33}) = (0.8, 0.6, 0.9)$. Our $target\_a_i = (0.5, 1, 0.5)$. Therefore, we find that there are two fit_a_is in the first set, similarly one and no fit_a_i for the second and the third set. Finally, we take the first set as the set of fit_a_i.

Step 4: *Breeding*

Now, the objective is to generate a new $child_\theta$ from $parent_\theta$. We choose recombination technique [11] as breeding technique. This new values are called two children $anew_i$ and $bnew_i$, where,

$anew_i = \alpha a_i + (1 - \alpha)b_i,$

$bnew_i = \beta b_i + (1 - \beta)a_i,$

where, α = random value between 0 to 1, and

β = random value between 0 to 1.

Step 5: *Joining*

We form the next generation parent by using new children. Joining equations are given as follows.

$a_i = anew_i$

$b_i = bnew_i$

Step 6: *Error Percentage of Participant Reliability*

We calculate the percentage of error of participant's reliability by dividing the total number of converged reliable nodes by the total number of reliable nodes.

The flow chart in Fig. 5 shows the summary of the procedure.

Now, we provide the formal algorithms of PBRE from Algorithms 1 to 7.

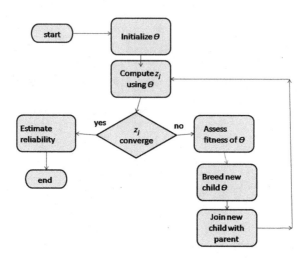

Fig. 5. The summary of the procedure.

Algorithm 1. Procedure PBRE

1: Take input: M, N, P, d, the observation matrix SC with random values either 0
   or 1
2: Initialize θ = random values between 0 and 1
3: Initialize $fit\_a(i)$ as NULL
4: Initialize $zcount=0$, z_j convergence metric
5: Calculate $target\_a(i) = \sum\limits_{i=1}^{M} \sum\limits_{j=1}^{N} \frac{SC(i,j)}{N}$
6: **while** z_j does not converge **do**
7:     **for** $i=1{:}M$ **do**
8:         $a(i, P + 1) = fit\_a(i)$ and add $fit\_a(i)$ in the population of $a(i)$
9:     **end for**
10:    **for** $j=1{:}N$ **do**
11:        **for** $K=1{:}P+1$ **do**
12:            $z(j, K)$
13:            **if** $z(j, K) = d$ **then**
14:                z_j = convergence counter
15:            **end if**
16:        **end for**
17:    **end for**
18:    Compute fitness using $Fit\_Parent()$ or $Replace\_Parent()$
19:    Apply $Breed()$, breeding
20:    Apply $Join()$, Joining
21:    Find reliability using $ReliabilityEstimation()$
22: **end while**

Algorithm 2. Computing probability that the event z_j is true or false: procedure $z(j, K)$

1: Calculate a_t =conditional probability that participant's observation is true given θ and event $z=1$.

2: $a_t(j, K) = \sum\limits_{i=1}^{M} a(i, K)^{SC(i,j)} (1 - a(i, K))^{(1-SC(i,j))}$

3: Calculate b_t =conditional probability that participant's observation is true given θ and event $z=0$.

4: $b_t(j, K) = \sum\limits_{i=1}^{M} b(i, K)^{SC(i,j)} (1 - b(i, K))^{(1-SC(i,j))}$

5: $z(j, K) = \frac{at1(j,K) \times d}{a_t 1(j,K) \times d + b_t 1(j,K) \times (1-d)}$

Algorithm 3. Procedure Fit_Parent()

1: This is to select closest a to $target\_a$ as fit_a
2: **for** $i = 1$ to M **do**
3:     **for** $K = 1$ to $P + 1$ **do**
4:         **if** $(0<= target\_a(i) <= 0.25$ AND $0<= a(i, K) <=0.25)$ OR $(0.25<target\_a(i) <= 0.75$ AND $0.25 <a(i, K) <=0.75)$ OR$(0.75<target\_a(i) <=1$ AND $0.25 <a(i, K) <=1)$ **then**
5:             $fit\_a(i)=a(i)$
6:         **end if**
7:     **end for**
8: **end for**
9: **return** fit_a

Algorithm 4. Procedure Replace_Parent()

1: This is to select closest set of a to set of $target\_a$ as fit_a
2: **for** $i = 1$ to M **do**
3:     **for** $K = 1$ to $P + 1$ **do**
4:         **if** $(0<= target\_a(i) <= 0.25$ AND $0<= a(i, K) <=0.25)$ OR $(0.25<target\_a(i) <= 0.75$ AND $0.25 <a(i, K) <=0.75)$ OR $(0.75<target\_a(i) <=1$ AND $0.25 <a(i, K) <=1)$ **then**
5:             $count(K) + +$
6:         **end if**
7:     **end for**
8: **end for**
9: $best=0$
10: **for** $K = 1$ to $P + 1$ **do**
11:     **if** $count(K)>best$ **then**
12:         $best =count(K)$
13:         $L=K$
14:     **end if**
15: **end for**
16: **for** $i=1$ to M **do**
17:     $fit\_a(i) = a(i, L)$
18: **end for**
19: **return** fit_a

Algorithm 5. Procedure breed()

1: **for** i=1 to M **do**
2:     **for** i=1 to P **do**
3:         $t(i, K) = \alpha \times a(i, K) + (1 - \alpha) \times b(i, K)$, α = between 0 to 1
4:         $s(i, K) = \beta \times b(i, K) + (1 - \beta) \times a(i, K)$, β = between 0 to 1
5:     **end for**
6: **end for**
7: **return** t, s

Algorithm 6. Procedure join()

1: Replace new children with parents
2: **for** $i = 1$ to M **do**
3:     **for** K=1 to P **do**
4:         $a(i, K) = t(i, K)$
5:         $b(i, K) = s(i, K)$
6:     **end for**
7: **end for**
8: **return** a, b

Algorithm 7. Procedure ReliabilityEstimation()

1: **for** i=1 to M **do**
2:     **if** $(0< $ $=$target_a(i)$<$ $=$ $0.25 AND 0<$ $=$fit_a(i,   K)$<$ $=$
    $0.25)OR(0.25<$target_a(i)$<$ $=$ $0.75 AND 0.25<$fit_a(i,   K)$<$ $=$
    $0.75)OR0.75<$target_a(i)$<=1 AND 0.25<$fit_a(i, K)$<=1$) **then**
3:         $truecount + +$, count of correct reliability estimation
4:     **end if**
5: **end for**
6: $error = (1 - \frac{truecount}{M}) \times 100$, percentage of error reliability estimation

---

4   Experimental Results

In this section, we present the experimental results to show the effectiveness of
PBRE method. We also compare our findings with the findings of another relevant algorithm Expectation Maximization [35]. The simulation of PBRE runs on
1.58 GHz *Intel Core 2 Duo Processor* with *2* GB memory. Simulation is done on
synthetic data sets where SC matrix is generated randomly. SC matrix contains
data of 1, 0 and if we used real time data set it would carry the similar property.
The performance metrics used to evaluate the methods are described as follows:

1. The *Error Percentage of Participant's Reliability* denotes the estimation of
   reliability of a participant to a converged event z.
2. The *Convergence Rate* denotes how quickly participant can report the correct event. It is computed by the participants' reliability divided by the total
   iteration needed to converge.

Now, we give the table of simulation parameters used for testing in Table 1.

Table 1. Simulation parameters

Parameters	Value
Participant number, M	30–900
Event number, N	2–10
Observation Matrix, SC	0,1
Probability of a reported TRUE event as TRUE, a	(0 to 1)
Probability of reported FALSE event as TRUE, b	(0 to 1)
Probability of the event C is actually TRUE, d	0.7
α	(0 to 1)
β	(0 to 1)

We carry out experiments using simulation to evaluate the performance of the proposed $PBRE$ scheme in terms of estimation accuracy of the probability that a participant is right or a measured variable is true compared to another existing reference method Expectation Maximization (EM) Method. We take the average of ten simulation runs. Variance we found is negligible. We consider two types of scenario. In a dense network, the number of participant nodes M is high.

Varying different parameters value, the error percentage was calculated. We now give the details as follows.

4.1 For Variable Number of Participants

We compare the estimation accuracy of PBRE (Fit_Parent and Replace_Parent) and Expectation Maximization(EM) scheme by varying the number of participants in the system.

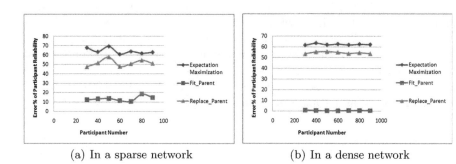

(a) In a sparse network   (b) In a dense network

Fig. 6. Error estimation for participant number

In Fig. 6(a), the number of participants is varied from 30 to 90. Two events and two sets of reliability per person are considered. We observe that, PBRE has a lower estimation error in participant reliability compared to EM scheme. Between two schemes of PBRE, Fit_Parent and Replace_Parent, Fit_Parent has much lower estimation error. This is because Fit_Parent takes only the fit values whereas Replace_Parent takes the fit set of values.

We run experiments for the increased number of participants from 300 to 900. The number in the set of reliability per person is 15. Event number is same as before e.i. two. Now, in Fig. 6(b), we observe that the error percentage decreases for Fit_Parent to 1 % which is compared to 10 to 15 % in Fig. 6 for participants with 4 sets of reliability per person. The reason behind this decline is due to the increased number in the set of reliability.

4.2 For Variable Number of Events

Here, we compare the results by varying the number of events from 2 to 10 for two cases.

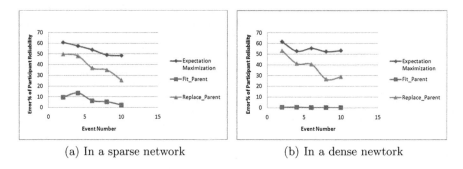

(a) In a sparse network   (b) In a dense newtork

Fig. 7. Error estimation for the events

In Fig. 7(a), experiments are run for a sparse network of 50 participants, 2–10 events and 4 set of reliability per person. Here also, PBRE shows better results than EM because, when the event number increases, $target\_a_i$ decreases (Line 6, Procedure PBRE). Therefore, there are more matches of a_i as fit_a_i to $target\_a_i$. We run experiments for the increased number of participants to 600 in Fig. 7(b). Here also, PBRE shows better results than EM because, when the event number increases, $target\_a_i$ decreases (Line 6, Procedure PBRE). Therefore, there are more matches of a_i as fit_a_i to $target\_a_i$.

4.3 Convergence Rate

We study the convergence vs. estimation accuracy of PBRE and EM scheme by varying the number of participants from 30 to 80. Event number is fixed at 2 and the set of reliability per person is at 4. In Fig. 8, we observe that the convergence rate for PBRE is a bit lower than the EM. Since PBRE has the lower error percentage of reliability than EM, it iterates more than EM to converge. Here, the convergence rate for Fit_Parent, Replace_Parent and EM are 0–2, 3.5–4.5 and 8–10 respectively.

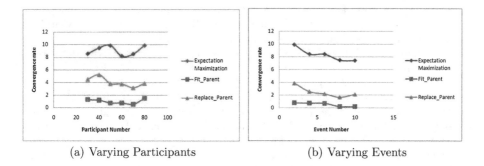

(a) Varying Participants   (b) Varying Events

Fig. 8. Convergence rate.

We then examine the results by varying number of events from 2 to 10. The Participant number is fixed at 50 and set of reliability per person is at 4. In Fig. 8(b), we find that the convergence rate for PBRE is lower than the EM. This is because of the similar fact happening in Fig. 8(a). Here, convergence rate for Fit_Parent, Replace_Parent and EM are 0–1, 2–4 and 7.5–10 respectively. We also observe that the rate in Fig. 8(a) is lower than the rate in Fig. 8(b) for increased number of events which is natural since the number of events more it takes more time to converge.

5   Conclusion

In this paper, we study the challenge of finding the node reliability in a participatory sensor network. We propose Population Based Reliability Estimation(PBRE). We computed conditional probability of event to be true with the given set of reliability. We use Genetic Algorithm to estimate the reliability by iterating fitness assessment, breeding and joining. We vary the number of participants, the number of events. The metrics for performance measurement are the error percentage of participantsŕeliable reports and the convergence rate. We compare the results with the result of another relevant and popular method Expectation Maximization for the truth finding. We find that our approach provides better results.

In future, we would like to provide some hybrid approach to have better results. Besides, we have a bias about ground truth more than 50 % probability. We would like to explore the impact of the uncertainty on our method.

References

1. Aggarwal, C., Abdelzaher, T.: Social sensing. In: Managing and Mining Sensor Data, pp. 237–297. Springer (2013)
2. Amintoosi, H., Allahbakhsh, M., Kanhere, S., Torshiz, M.N.: Trust assessment in social participatory networks. In: International eConference on Computer and Knowledge Engineering (ICCKE), pp. 437–442 (2013)
3. Burke, J., Estrin, D., Hansen, M., Parker, A., Ramanathan, N., Reddy, S., Srivastava, M.B.: Participatory sensing. In: World-Sensor-Web: Mobile Device Centric Sensor Networks and Applications, pp. 117–134 (2006)
4. Choudhury, T., Philipose, M., Wyatt, D., Lester, J.: Towards activity databases: using sensors and statistical models to summarize peoples lives. IEEE Data Eng. Bull. **29**(1), 49–58 (2006)
5. Cox, L.P.: Truth in crowd sourcing. IEEE Secur. Priv. **9**(5), 74–76 (2011)
6. Deng, L., Cox, L.P.: Livecompare: Grocery bargain hunting through participatory sensing. In: The Workshop on Mobile Computing Systems and Applications (HotMobile), NY, USA, pp. 1–6. ACM (2009)
7. Dong, X.L., Berti-Equille, L., Srivastava, D.: Integrating conflicting data: the role of source dependence. VLDB Endow. **2**(1), 550–561 (2009)
8. Dong, X.L., Berti-Equille, L., Srivastava, D.: Truth discovery and copying detection in a dynamic world. VLDB Endow. **2**(1), 562–573 (2009)
9. Dua, A., Bulusu, N., Feng, W.-C., Hu, W.: Towards trustworthy participatory sensing. In: USENIX Conference on Hot Topics in Security (HotSec), CA, USA (2009)
10. Eagle, N., Pentland, A.S., Lazer, D.: Inferring friendship network structure by using mobile phone data. Natl. Acad. Sci. **106**(36), 15274–15278 (2009)
11. Eiben, A.E., Smith, J.E.: Introduction to Evolutionary Computing. Springer, Heidelberg (2003)
12. Eisenman, S.B., Miluzzo, E., Lane, N.D., Peterson, R.A., Ahn, G.-S., Campbell, A.T.: The bikenet mobile sensing system for cyclist experience mapping. In: International Conference on Embedded Networked Sensor Systems (SenSys), pp. 87–101. ACM (2007)
13. Estrin, D.L.: Participatory sensing: applications and architecture. In: ACM MobiSys, pp. 3–4 (2010)
14. Gilbert, P., Cox, L.P., Jung, J., Wetherall, D.: Toward trustworthy mobile sensing. In: Computing Systems and Applications, NY, USA, pp. 31–36 (2010)
15. Gilbert, P., Jung, J., Lee, K., Qin, H., Sharkey, D., Sheth, A., Cox, L.P.: YouProve: authenticity and fidelity in mobile sensing. In: International Conference on Embedded Networked Sensor Systems (SenSys), NY, USA, pp. 176–189 (2011)
16. Hicks, J., Ramanathan, N., Kim, D., Monibi, M., Selsky, J., Hansen, M., Estrin, D.: Wellness: an open mobile system for activity and experience sampling. In: Wireless Health (WH), pp. 34–43. ACM (2010)
17. Lane, N., Choudhury, T., Campbell, A.: Bewell: a smart phone application to monitor, model and promote well-being. In: International ICST Conference on Pervasive Computing Technologies for Healthcare (2011)

18. Le, H.K., Pasternack, J., Ahmadi, H., Gupta, M., Sun, Y., Abdelzaher, T.F., Han, J., Roth, D., Szymanski, B.K., Adali, S.: Apollo: towards fact finding in participatory sensing. In: ACM IPSN, pp. 129–130 (2011)

19. Lenders, V., Koukoumidis, E., Zhang, P., Martonosi, M.: Location-based trust for mobile user-generated content: applications, challenges and implementations. In: ACM Workshop on Mobile Computing Systems and Applications (HotMobile), NY, USA, pp. 60–64 (2008)

20. Lu, H., Frauendorfer, D., Rabbi, M., Mast, M.S., Chittaranjan, G.T., Campbell, A.T., Gatica-Perez, D., Choudhury, T. Stress sense: detecting stress in unconstrained acoustic environments using smart phones. In: ACM Conference on Ubiquitous Computing (UbiComp), pp. 351–360 (2012)

21. Madan, A., Cebrin, M., Moturu, S.T., Farrahi, K., Pentland, A.: Sensing the health state of a community. IEEE Pervasive Comput. **11**(4), 36–45 (2012)

22. Madan, A., Moturu, S.T., Lazer, D., Pentland, A.S.: Social sensing: obesity, unhealthy eating and exercise in face-to-face networks. In: Wireless Health (WH), pp. 104–110 (2010)

23. Olgun, D.O., Gloor, P.A., Pentland, A.: Wearable sensors for pervasive healthcare management. In: Pervasive Health, pp. 1–4 (2009)

24. Oliveira, M.P.G., Medeiros, E.B., Davis, C.A.: Planning the acoustic urban environment: a gis centered approach. In: GIS, pp. 128–133 (1999)

25. Reddy, S., Shilton, K., Denisov, G., Cenizal, C., Estrin, D., Srivastava, M.: Biketastic: sensing and mapping for better biking. In: ACM Conference on Human Computer Interaction (CHI), pp. 1817–1820 (2010)

26. Roy, D., et al.: The human speechome project. In: Vogt, P., Sugita, Y., Tuci, E., Nehaniv, C.L. (eds.) EELC 2006. LNCS (LNAI), vol. 4211, pp. 192–196. Springer, Heidelberg (2006)

27. Ryder, J., Longstaff, B., Reddy, S., Estrin, D.: Ambulation: A tool for monitoring mobility patterns over time using mobile phones. In: CSE, vol. 4, pp. 927–931. IEEE (2009)

28. Srivastava, M., Abdelzaher, T., Szymanski, B.: Human-centric sensing. Philos. Trans. R. Soc. A Math. Phys. Eng. Sci. **370**(1958), 176–197 (2012)

29. Tang, L.A., Yu, X., Kim, S., Han, J., Hung, C.-C., Peng, W.C.: Tru-alarm: Trustworthiness analysis of sensor networks in cyber-physical systems. In: International Conference of Data Management (ICDM), pp. 1079–1084 (2010)

30. Tilak, S.: Real-world deployments of participatory sensing applications: current trends and future directions. ISRN Sens. Netw. **2013**, 8 (2013)

31. Wang, C., Burris, M.A., Ping, X.Y.: Chinese village women as visual anthropologists: a participatory approach to reaching policy makers. Soc. Sci. Med. **42**(10), 1391–1400 (1996)

32. Wang, D., Abdelzaher, T., Ahmadi, H., Pasternack, J., Roth, D., Gupta, M., Han, J., Fatemieh, O., Le, H., Aggarwal, C.: On bayesian interpretation of fact-finding in information networks. In: Information Fusion (FUSION), pp. 1–8 (2011)

33. Wang, D., Abdelzaher, T., Kaplan, L., Aggarwal, C.: On quantifying the accuracy of maximum likelihood estimation of participant reliability in social sensing (2011)

34. Wang, D., Abdelzaher, T., Kaplan, L., Aggarwal, C.: Recursive fact-finding: a streaming approach to truth estimation in crowd sourcing applications. In: International Conference on Distributed Computing Systems (ICDCS) (2013)

35. Wang, D., Kaplan, L., Le, H., Abdelzaher, T.: On truth discovery in social sensing: a maximum likelihood estimation approach. In: ACM International Conference on Information Processing in Sensor Networks (IPSN), pp. 233–244 (2012)
36. Yin, X., Han, J., Yu, P.S.: Truth discovery with multiple conflicting information providers on the web. In: International Conference on Knowledge Discovery and Data Mining (KDD), New York, USA, pp. 1048–1052 (2007)
37. Yin, X., Tan, W.: Semi-supervised truth discovery. In: ACM International Conference on World Wide Web, New York, USA, pp. 217–226 (2011)

Knowledge Acquisition and Applications

Knowledge Acquisition for Learning Analytics: Comparing Teacher-Derived, Algorithm-Derived, and Hybrid Models in the Moodle Engagement Analytics Plugin

Danny Y.T. Liu[1,2(✉)], Deborah Richards[2], Phillip Dawson[3],
Jean-Christophe Froissard[4], and Amara Atif[2]

[1] Educational Innovation Team, The University of Sydney, Sydney, Australia
danny.liu@sydney.edu.au
[2] Faculty of Science and Engineering, Macquarie University, Sydney, Australia
{danny.liu,deborah.richards,amara.atif}@mq.edu.au
[3] Centre for Research in Assessment and Digital Learning,
Deakin University, Melbourne, Australia
p.dawson@deakin.edu.au
[4] Faculty of Arts, Macquarie University, Sydney, Australia
chris.froissard@mq.edu.au

Abstract. One of the promises of big data in higher education (learning analytics) is being able to accurately identify and assist students who may not be engaging as expected. These expectations, distilled into parameters for learning analytics tools, can be determined by human teacher experts or by algorithms themselves. However, there has been little work done to compare the power of knowledge models acquired from teachers and from algorithms. In the context of an open source learning analytics tool, the Moodle Engagement Analytics Plugin, we examined the ability of teacher-derived models to accurately predict student engagement and performance, compared to models derived from algorithms, as well as hybrid models. Our preliminary findings, reported here, provided evidence for the fallibility and strength of teacher- and algorithm-derived models, respectively, and highlighted the benefits of a hybrid approach to model- and knowledge-generation for learning analytics. A human in the loop solution is therefore suggested as a possible optimal approach.

Keywords: Learning analytics · Goal seeking · Human and machine models · Hybrid models · Knowledge acquisition

1 Introduction

In the era of Big Data where knowledge is machine generated from raw data, one might assume that the knowledge acquisition bottleneck identified in 1980 by Feigenbaum [1] that relies on humans to articulate their knowledge has been overcome. This bottleneck was seen to be the major impediment to the advancement of artificial intelligence and highlighted the difficulties associated with expressing and representing what human experts "know" [1]. Challenges included gaining access to domain experts willing and

H. Ohwada and K. Yoshida (Eds.): PKAW 2016, LNAI 9806, pp. 183–197, 2016.
DOI: 10.1007/978-3-319-42706-5_14

able to articulate their knowledge and, if this hurdle was overcome, having to understand and encode the terms and concepts in often unfamiliar domains and find a way to validate and maintain the resultant knowledge base(s) [2].

Data-driven approaches to knowledge discovery may seem to circumvent the knowledge engineering process and the need for knowledge acquisition from humans, but even advanced data processing methods or algorithms often include one or more steps that rely on human input, for example to set weights or identify cases and features. Even where human input is not essential and machine-learnt weights and features can be used, inclusion of humans at key points in the process can significantly reduce the search space and improve performance [3]. As with any computer software, verification can be done by a machine, but validation requires human involvement to confirm that it is fit for purpose and addresses needs. Similarly, measures of interestingness and other machine-derived indicators will not be able to replace the domain expert when it comes to validation and utilization of the knowledge uncovered. We believe this is particularly true in the subfield of learning analytics, where the activities undertaken by learners have been designed by the domain expert (teacher) to aid learning and thus measurement of the learner's performance in these activities is fundamentally tied to the domain knowledge and pedagogical strategy (or problem-solving knowledge) of the teacher [4].

Together with other researchers, we have developed a tool, known as the Moodle Engagement Analytics Plugin (MEAP), that uses data mining methods on student data associated with their use of the popular Moodle learning management system (LMS), to identify levels of student engagement in a course and determine a risk rating that could predict success or failure in a course [5–7]. To be able to calculate the risk rating it is necessary to acquire course-related knowledge from the teacher in the form of the salient features (or triggers) that indicate engagement and participation, and the thresholds (or parameters) and weightings of each of these features that may indicate risk. Similar to the knowledge acquisition bottleneck, teachers have found providing these parameters and weightings a barrier to adoption of the system because assigning them is time-consuming and hard to articulate and validate, leading to a lack of confidence in the risk rating produced.

To address this barrier, in this paper we present a preliminary exploration of the use of machine learning on historical data to generate the parameters and weightings. We correlate the results of the human teacher-derived models with machine (algorithm)-derived models against actual student performance data to determine which performs better. We also investigate the performance of 'hybrid' models.

In the next section we introduce the architecture of MEAP. In Sect. 3 we present our methodology including the courses chosen for our evaluation, how teachers assigned parameters and weightings, the algorithm used to determine these, and how to compare the different models. Results appear in Sect. 4, followed by discussion in Sect. 5, and conclusions and future directions in Sect. 6.

2   The Moodle Engagement Analytics Plugin

The Moodle Engagement Analytics Plugin is open source software that plugs into the Moodle LMS[1]. It is built around 'indicators' that each address an aspect of students' expected engagement with a course. The three primary indicators are assessment, forums, and logins, in keeping with literature that highlights these metrics as informative for student engagement and performance [8–10]. These indicators read data from the Moodle database and, based on user-defined parameters and weightings, are able to calculate a risk rating for each indicator for each student in a course (Fig. 1). These risk ratings for the individual indicators are then weighted by user-definable weightings, and form a total risk rating for each student which is then reported by the tool (Fig. 1). Previous work has validated the efficacy of the total risk rating to reflect student course performance, and therefore provide a useful measure of student engagement and disengagement [6].

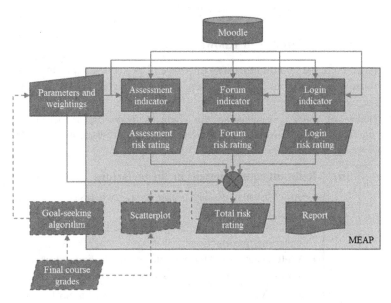

Fig. 1. Architecture of the Moodle Engagement Analytics Plugin (MEAP). Additional structures added for this study indicated by dashed outlines.

The original release of MEAP allowed teachers to self-define the parameters and weightings for the tool, ostensibly using their domain knowledge and experience. To enable the current study and allow teachers to leverage machine knowledge derived from the actual Moodle dataset, we added functionality into MEAP that could algorithmically determine optimal parameters and weightings using these data (Fig. 1).

---

[1] Available at https://github.com/netspotau/moodle-report_engagement as open source software.

What follows is a quantitative analysis of such teacher-derived and algorithm-derived knowledge in the context of this learning analytics tool.

3   Methodology

3.1   Course Selection

We present a preliminary examination of three undergraduate courses taught at Macquarie University. Each course was delivered in 2014, and again in 2015. There were two humanities courses (first year, HUM1, and second year, HUM2) and one first year science course (SCI1) (course codes have been deidentified). These courses were examined due to their relatively high failure rates (15-26 %) and sizeable classes (Table 1), making them important for analysis [6]. This study was approved by the Macquarie University Human Research Ethics Committee (approval numbers 5201300866 and 5201500031). Table 1 provides a summary of each course's relevant Moodle learning designs, with a focus on aspects most relevant (and accessible) to MEAP.

3.2   Teacher-Derived Models of Knowledge

One teacher (the course coordinator) was interviewed per course early in the semester that the course was offered in 2015. Teachers were first given a basic introduction to

Table 1. Outline of courses examined in this study. Activities falling outside the study timeframe (up until the end of week four; see Sect. 3.4) or not captured online are denoted in square brackets

Course	Size	Relevant course designs in 2014 offerings
HUM1	151	• Assessments: weekly online quiz, [two essays, pre-readings, class participation] • Forums: announcements, general discussion, welcome/introductions • Well-structured Moodle site with readings, links, lecture slides and recordings, assessment guidance
SCI1	57	• Assessments: weekly online exercise, diagnostic quiz, assignment, [two assignments, test, and exam] • Forums: announcements and general discussion • Well-populated Moodle site with lecture and tutorial notes, assessment submission boxes
HUM2	78	• Assessments: online discussion, [class participation, essay, exam] • Forums: announcements, general discussion, weekly discussion (assessed) • Well-structured Moodle site with weekly readings and additional resources, links, lecture notes and recordings, assignment submission boxes

MEAP, which involved the researchers describing the data from the Moodle logs that were analyzed by MEAP, the three indicators (assessment, forum, and login), and how the indicators each produced a risk rating that were combined to form the total risk rating. Teachers were then asked to set the parameters for each indicator and the indicator weightings, by conceptualizing what they expected of a good student. They were also interviewed on their conceptions of identifying student engagement (e.g. what might be effective variables to measure engagement and performance), and any challenges they perceived in using the tool.

3.3   Algorithm-Derived Models of Knowledge

Algorithm design. We developed a simple goal-seeking algorithm and embedded it into MEAP. The algorithm was loosely based on a simulated annealing approach [11], where parameters would be iteratively adjusted to improve the outcome of an acceptance function, which was, in this case, maximizing the inverse correlation between total risk rating and final course grade for each student (i.e. a correlation closer to -1 was preferred, as total risk rating should be inversely correlated with final course grade, used here as a proxy for student performance). Pseudocode for the goal-seeking algorithm is presented below (Algorithm 1), and the full code is open source[2].

```
Set array of parameters with starting values
For i = 1 through i_max
   Randomly shuffle array of parameters
   For each parameter
      For j = j_max through 1
         factor = random number between 1.2 and (j * 1.5)
         new_hi = current_value * factor
         new_lo = current_value / factor
         For each value (new_hi, new_lo, current_value)
            Update indicator settings
            Calculate risk ratings for each student
            Correlate risk ratings with outcome variable
         Select value that yields best correlation
            Set current_value to be that value
Visually report best correlation to user
```

Algorithm 1. Pseudocode for goal-seeking algorithm used to determine optimal parameters.

The algorithm starts with pre-defined starting values and, at each iteration, uses a pseudorandom factor to move each parameter higher and lower, and tests these parameter values to find the best direction to move in to improve the correlation

---

[2] https://github.com/dannyliu-mq/moodle-mod_engagement/tree/indicator_helper and https://github.com/dannyliu-mq/moodle-report_engagement/tree/indicator_helper.

between risk rating and the outcome variable (in our case, the final course grade was used as a proxy for student performance). The algorithm was crudely designed so that at each iteration of j, the movement would become more limited as the algorithm approached an optimal solution. Our aim in embedding this goal-seeking algorithm into the existing Moodle plugin was so that teachers could ultimately use it themselves to assist in determining optimal parameters and weightings. This was presented as a graphical user interface (Fig. 2). However, being embedded within a Moodle page imposed some technical limitations on this approach. Most notably, this included a PHP script execution timeout limit (typically 30 s) which, in our technological context, practically constrained the number of steps through which the algorithm could iterate. We are currently working on an alternate approach that uses client-side asynchronous requests to the server for each iteration, mitigating these timeout limitations, in conjunction with a genetic algorithm instead of simulated annealing so that the 28 parameters and weightings can be searched simultaneously.

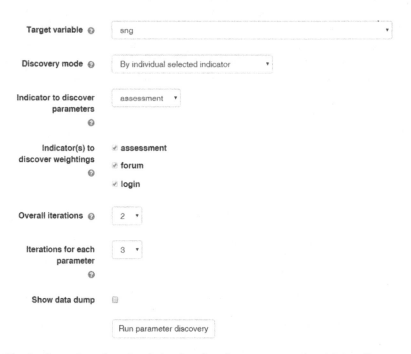

Fig. 2. Screenshot of teacher-facing interface for parameter and weighting discovery

Deriving parameters using the goal-seeking algorithm. To determine the algorithm-derived parameters and weightings for each course, the target outcome was specified as course final grade and the number of steps (i and j; Algorithm 1) were typically set to 4-5. The algorithm was first run on each indicator (assessment, forum, and login) separately to determine the optimal parameters for each indicator. The algorithm was then run to determine the optimal weightings of each of the three indicators. Finally, the

Pearson correlation coefficient was reported by the tool, expressing the correlation between the total risk rating and final course grade for each student in the course.

3.4   Comparing Teacher- and Algorithm-Derived Models

To determine the correlation between total risk ratings reported by MEAP and final course grade for each student, the parameters and weightings that each teacher determined at the beginning of semester one, 2015, for their course were entered into MEAP after the conclusion of the semester and final course grades were available. The new functionality built into MEAP allowed the reporting of the Pearson correlation coefficient in each course between each students' course final grade and calculated risk ratings. These correlations were calculated for each separate indicator (assessment, forum, and login), as well as for the total risk rating.

The same method was followed to determine the correlations between course final grades and risk ratings as calculated according to algorithm-derived parameters and weightings. An overview of this process is presented in Fig. 3. Because previous work on MEAP identified week four (out of a 13-week semester) as a best compromise between early detection and the availability of sufficient data [6], we limited MEAP to only use Moodle data available from the beginning of semester up until the end of week four, in each of the offerings of the courses analyzed.

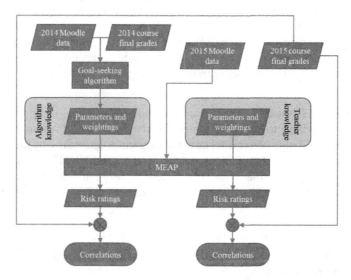

Fig. 3. Overview of knowledge discovery and application process. Algorithm knowledge was derived through goal-seeking analysis of previous (2014) course offering data, while teacher knowledge was derived from human understanding of each course. Both models were tested by generating risk ratings through the Moodle Engagement Analytics Plugin (MEAP) and correlating these with course data from 2015.

4   Results

4.1   Teacher- and Algorithm-Derived Models

The parameters for each individual indicator (assessment, forum, and login), as well as the overall weightings of each indicator, are presented in Table 2 as determined by the teacher of each course, and by the goal-seeking algorithm within each course. In some situations, parameters and weightings were differentially ignored by teachers or the algorithm; that is, their weighting (or importance) was set to 0 % and therefore had no impact upon risk rating calculation.

Teacher-derived models. For HUM1, the teacher's conception of what was important in the course impacted the settings they determined, although they acknowledged that this may differ to how students perceive the course: "*At a glance I can basically already highlight, for instance, that level of engagement and basically whether they're logging into [the LMS]... I've given you certain weightings on what I think is important for them but maybe my expectations of them might not be realistic*". This was reflected in their preference for the login indicator, and the length of each session (Table 2).

Similarly, the teacher of SCI1 had firm beliefs regarding the prevailing indicator, in this case assessment: "*I think participation in - or I should say submission of the assessment tasks... This is quite an effective way of measuring performance.*" Again, this was reflected in their chosen weightings (preferring the assessment indicator), and the lack of leeway given to late submissions (Table 2).

For HUM2, the teacher specified that "*the online discussion is crucial and so if students aren't involved on the online discussion early on they fall behind very quickly in terms of understanding concepts.*" Perhaps reflecting the nature of the discussions being online (and therefore students needing to log in to access them), this teacher more evenly balanced the forum and login indicators and placed emphasis on new posts and reading posts (Table 2).

Algorithm-derived models. The parameters and weightings determined by the goal-seeking algorithm to optimize the correlation coefficient are also presented in Table 2. There was a stark contrast between the weightings determined by the teacher and those determined by the algorithm, especially in SCI1 and HUM2. Indicator-specific parameters were also variable, with little agreement between teacher- and algorithm-derived models. There was also no appreciable pattern of difference between these two models.

4.2   Predictive Power of Teacher-Derived, Algorithm-Derived, and Hybrid Models

The correlations between final course grades and risk ratings reflected this dissonance between teacher- and algorithm-derived models. The scatterplots in Fig. 4 show the correlation between final course grade for each student and their corresponding risk rating for a particular indicator (login) in HUM1. Although the teacher-derived model (generated in 2015 and applied to the 2015 course offering) is correlated in the right

Table 2. Parameters and weightings determined by teachers (T) and the algorithm (A) for the three courses examined in this study. Dashes indicate ignored parameters or weightings

Variable		HUM1		SCI1		HUM2	
		T	A	T	A	T	A
Overall indicator weightings	Assessment (%)	30	14	60	–	15	1
	Forum (%)	10	7	10	100	45	89
	Login (%)	60	79	30	–	40	10
Assessment indicator	Grace days	0	2	0	6.4	0	0
	Maximum days	5	3.2	0	7	5	14
	Submitted weighting (%)	50	50	50	0	50	50
	Non-submitted weighting (%)	100	39	100	100	100	100
Forum indicator	New posts per week (no risk, max risk, % weighting)	1, 0, 40	–	0.5, 0, 12	2, 0, 27	1, 0, 40	0.05, 0, 26
	Read posts per week (no risk, max risk, % weighting)	3, 0, 40	0.6, 0, 77	1, 0, 12	0.8, 0, 46	1, 0, 40	4.3, 0, 38
	Replies per week (no risk, max risk, % weighting)	3, 0, 30	0.3, 0, 9	1, 0, 20	–	1, 0, 20	0.5, 0, 23
	Total posts per week (no risk, max risk, % weighting)	–	0.6, 0, 14	1, 0, 56	2, 0, 27	–	0.3, 0, 13
Login indicator	Logins in the past week (number, % weighting)	4, 15	34, 29	–	4, 32	3, 25	4, 11
	Logins per week (number, % weighting)	4, 15	6, 36	2, 50	13, 15	3, 25	2.3, 66
	Length of login (minutes, % weighting)	20, 60	5, 35	30, 50	3.8, 2	20, 25	53, 23
	Time since last login (days, % weighting)	2, 10	–	–	14, 51	7, 25	–
	Session length (minutes)	20	60	60	60	20	60

(negative) direction, it is non-significant and has a low effect size [12]. In comparison, the algorithm-derived model (generated from 2014 data and applied to the 2015 course offering) had a medium-large effect size and was highly significant.

In all three courses examined, the algorithm-derived model outperformed the teacher-derived model when the overall risk rating was correlated with final course grade (Table 3 and Fig. 5). This pattern was also reflected for the login indicator, where the correlation coefficients were consistently closer to -1 in the algorithm-derived models (Table 3). In fact, the teacher-derived models for the login indicator in SCI1 and HUM2 led to positive correlation coefficients, meaning that the indicator

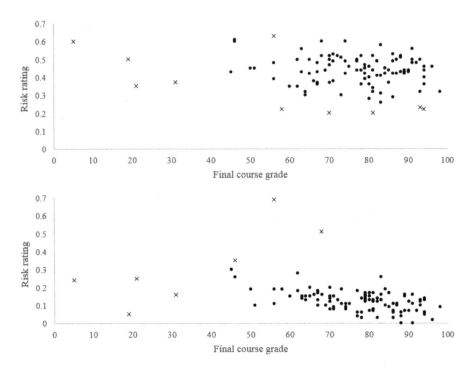

Fig. 4. Correlations between risk rating and course final grade based on human-derived (top) and machine-derived (bottom) parameters for the login indicator in HUM1. Crossed points were considered outliers (Z-score greater than 2). Correlation coefficients are −0.1001 (top, p > 0.30) and −0.4271 (bottom, p < 0.0001).

would be associating high-performing students with higher risk ratings. The higher power of the algorithm-derived models was also seen for the assessment indicator, although not to the same extent (Table 3). The zero correlation coefficient in SCI1 for the assessment indicator was a product of the calculations that MEAP performs and the teacher-derived setting of '0' for 'maximum days' (Table 2). This actually nullified the assessment risk calculation due to the internal algorithmic design of the assessment indicator, which scales the risk calculation from zero days to the 'maximum days' setting. This possibly reflects a misunderstanding by the teacher of the somewhat opaque underlying mechanisms of risk rating calculation. In terms of the forum indicator, the teacher- and algorithm-derived models had similar power (Table 3).

To examine the power of a crude hybrid model, we took the mean of teacher- and algorithm-derived parameters and weightings and used these to update the MEAP settings and calculate risk ratings, and then the correlations between these and course final grades. Interestingly, in some instances (HUM1) the hybrid model underperformed the algorithm-derived model, but in SCI1 and HUM2 the hybrid model outperformed both teacher- and algorithm-derived models (Table 3 and Fig. 5).

Table 3. Correlations between course final grades and risk ratings (for individual indicators and overall) as determined from teacher, algorithm, and hybrid models. * p < 0.5; ** p < 0.01; *** p < 0.001. Note: no assessments were detected in HUM2 during the study time period, hence these data were not available (N/A).

		Correlation coefficients			
		Assessment risk rating	Forum risk rating	Login risk rating	Overall risk rating
HUM1	Teacher	−0.493 ***	−0.438 ***	−0.100	−0.172
	Algorithm	−0.494 ***	−0.398 ***	−0.427 ***	−0.379 ***
	Hybrid	−0.494 ***	−0.345 ***	−0.254 **	−0.260 **
SCI1	Teacher	0.000	−0.172	+0.377 ***	+0.300 **
	Algorithm	−0.375 ***	−0.137	−0.031	−0.137
	Hybrid	−0.385 ***	−0.176	−0.032	−0.375 ***
HUM2	Teacher	N/A	−0.245 *	+0.123	−0.224 *
	Algorithm	N/A	−0.276 *	−0.149	−0.247 *
	Hybrid	N/A	−0.239 *	−0.196	−0.326 **

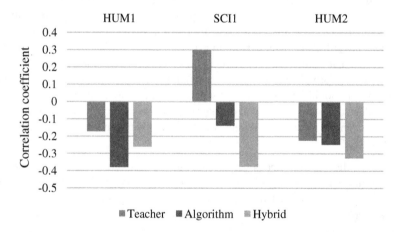

Fig. 5. Pearson correlation coefficients between overall risk ratings and course outcome, comparing the coefficients derived from teacher- and algorithm-derived models and hybrid models. A correlation coefficient closer to −1 indicates a better correlation. (Color figure online)

5   Discussion

The hybrid model can be seen as a rudimentary human in the loop solution. Human in the loop learning can take the form of active learning or learning via queries [13], where the query drives the interaction. Alternatively, the human can drive the interaction as in the case of intelligent user interfaces or personalization agents, where the human is able to drive the learning while the system observes the human's behavior [14].

Returning to our earlier consideration of whether the knowledge acquisition bottleneck remains a problem, and focusing within the context of learning analytics on student data captured in an LMS, we note that reliance on the human teacher to provide the parameters and weightings can be a major problem. Not only did the teachers in our study report that this was a barrier to usage for them because of the difficulty of coming up with initial parameters and weightings, it was also time-consuming to explore and make sense of the resultant risk ratings and optimize the settings. While in our case the domain expert, rather than a knowledge engineer, was responsible for encoding their own knowledge of their course and their students in the form of assigning indicator parameters and weightings, they faced similar problems to those faced by knowledge engineers. In addition to facing issues similar to limited availability and inability to articulate what they knew, it could also be said that they experienced other reported contributing factors to the knowledge acquisition bottleneck. Ruqian [2] enumerates the following problems faced by the knowledge engineer:

1. They must "extract as much knowledge as possible from the expert's memory and behavior; what the expert provides is only raw material, often mixed with personal biases, even wrong conclusions; this imply the need to screen out, test and reorganize the knowledge obtained from the expert;
2. knowledge is not equal to experience; experience is not always representable; it may be fuzzy and inconsistent; it may appear in the form of inspiration and randomly emerging ideas; the expert has difficulty in explaining it and the knowledge engineer has difficulty in understanding it;
3. there is no clear border between domain knowledge and common sense knowledge; the latter is informal, infinite, continuous and exists everywhere; it is difficult to decide what should be acquired and what should not be acquired;
4. knowledge cannot be acquired at one stroke; it has to be accumulated during a long process; even the most experienced expert is not able to provide this knowledge at a stretch." (p. 2)

Similarly, the teacher needs experience to be able to assign parameters and weightings and make sense of the results. However, just as it has taken time to acquire that knowledge and the knowledge itself is evolving, these settings are mutable and the assignment process is iterative and time-consuming. Moreover, their experience does not necessarily translate into settings that are useful or accurate. This was particularly visible in SCI1 where the teacher-derived model resulted in a positive correlation between total risk rating and course final grade, implying that the teacher-derived model actually reflected the engagement expected from poor performing students, not high performing ones. In this course for this teacher, an additional complicating factor was a misunderstanding of the impact of certain parameters (notably for the assessment indicator) that further contributed to poor performance of the teacher-derived model. This is related to the idea of model 'comprehensibility' [15], where calculation methods that are relatively opaque to the end user (in this case, the teacher) can obscure the usefulness of learning analytics approaches. Further, it could be said that the teacher's mental model of the risk rating calculation was flawed, resulting in an unfair (at best) or invalid (at worst) comparison between algorithm and teacher.

Although the three teacher-derived models investigated in this preliminary study consistently underperformed, the algorithm-derived models were not consistent outperformers as may be expected from a goal-seeking approach. At one level, this may be a reflection of suboptimal algorithm choice and design as well as limitations with the technology platform; indeed, genetic algorithms and related evolutionary approaches may be better suited to the optimization problem presented in the MEAP parameters and weightings [16]. However, the inconsistent outperformance of algorithm-derived models may also suggest that data-driven acquisition of knowledge by machines (at least in this instance) may not be completely adequate, and that some human input is necessary. Indeed, the simple hybrid models tested were the best performing in two of the three cases. To our knowledge, this is the first report of a learning analytics approach where human and machine models of student engagement are compared and hybridized, and suggests the importance of expert knowledge supported by data-driven knowledge acquisition processes. This provides some preliminary but cautionary evidence against the preponderance of large-scale learning analytics approaches that rely on purely machine-derived models of student engagement and performance [10, 17, 18], and exemplifies the symbiotic roles of human and machine in knowledge acquisition for learning analytics. A key future research direction for developing human in the loop hybrid models will be to move beyond simply taking the mean of teacher- and algorithm-derived models towards having teachers fine-tuning and adapting algorithm-derived models.

Our findings also provide further supporting evidence for a growing perspective in learning analytics that there is no one model that is suitable for all courses [19]. That is, the knowledge surrounding a course is unique and related to instructional and other contexts. The diversity of teacher-derived models reflects this, as do the range of parameters and weightings determined by the algorithm as best-fitting for each course. We suggest that the analysis we have done, and the correlation and parameter/ weighting discovery tool that we have built, could potentially aid teachers to revisit their assumptions and knowledge, and potentially modify their teaching strategies and learning designs accordingly. For example, a teacher could, *de novo*, determine their own parameters and weightings and then see how these impact the correlation between risk rating and final course grade for a previous course offering. Then, they could run the discovery engine and derive parameters and weightings using the goal-seeking algorithm. In this extended knowledge acquisition process, they could then trial hybrid models that combine machine-suggested settings with their domain knowledge and experience and settle on a set of parameters and weightings that optimize the correlation between risk rating and course final grade. They would then apply this to the current course offering, and subsequently apply this iterative cycle to accumulate knowledge over time.

6 Conclusions and Future Directions

We have provided preliminary evidence that suggests expert teacher knowledge for learning analytics can sometimes be outperformed by knowledge derived by data-mining algorithms, and that a hybrid approach may be optimal in some instances. Specifically, we have built such knowledge discovery mechanisms into MEAP, an open source learning analytics plugin for the Moodle LMS. Our results also support the growing trend in learning analytics research that emphasizes knowledge of instructional and other contexts in building accurate models.

As future work, a fully-integrated human in the loop approach that provides an intelligent and adaptive user interface able to guide the teacher in setting parameters and weightings (perhaps via seeding of initial settings if historical data from previous course offerings exist and by visualizing 'what-if' scenarios using alternative settings) would accelerate the task of determining optimal models. If historical data are not available, the settings could be seeded based on courses with similar learning designs or other characteristics. This is similar to a case-based reasoning approach where experience from other contexts is applied to knowledge-based systems [20]. Alternatively, an improved algorithmic approach (such as genetic algorithms) may provide more suitable seed settings for such exploration.

Acknowledgements. This work was supported by a Macquarie University Teaching Development Grant. We wish to thank all the staff who worked with us on this study, and our colleagues for many insightful conversations. Initial development and conceptualization of MEAP was supported by a NetSpot Innovation Fund grant, and conducted by a team including Adam Olley, Ashley Holman, Angela Carbone, and others.

References

1. Feigenbaum, E.A.: Knowledge engineering: the applied side of artificial intelligence. Ann. New York Acad. Sci. **426**, 91–107 (1984)
2. Ruqian, L.: New Approaches to Knowledge Acquisition. World Scientific Series in Computer Science. World Scientific Pub. Co. Inc. (1994)
3. Xu, H., Hoffmann, A.: RDRCE: combining machine learning and knowledge acquisition. In: Kang, B.-H., Richards, D. (eds.) PKAW 2010. LNCS, vol. 6232, pp. 165–179. Springer, Heidelberg (2010)
4. Lockyer, L., Heathcote, E., Dawson, S.: Informing pedagogical action: aligning learning analytics with learning design. Am. Behav. Sci. **57**, 1439–1459 (2013)
5. Liu, D.Y.T., Froissard, J.-C., Richards, D., Atif, A.: An enhanced learning analytics plugin for Moodle: student engagement and personalised intervention. In: Reiners, T., von Konsky, B.R., Gibson, D., Chang, V., Irving, L., Clarke, K. (eds.) 32nd Conference of the Australasian Society for Computers in Learning in Tertiary Education, Perth (2015)
6. Liu, D.Y.T., Froissard, J.-C., Richards, D., Atif, A.: Validating the effectiveness of the moodle engagement analytics plugin to predict student academic performance. In: 2015 Americas Conference on Information Systems, Puerto Rico (2015)
7. Dawson, P.: Analytics block to identify students at risk of disengaging. Paper presented at 2012 Moodle Moot, Gold Coast, Queensland, Australia (2012)

8. Macfadyen, L.P., Dawson, S.: Mining LMS data to develop an "early warning system" for educators: a proof of concept. Comput. Educ. **54**, 588–599 (2010)

9. Romero, C., Ventura, S., García, E.: Data mining in course management systems: Moodle case study and tutorial. Comput. Educ. **51**, 368–384 (2008)

10. Jayaprakash, S.M., Moody, E.W., Lauría, E.J.M., Regan, J.R., Baron, J.D.: Early alert of academically at-risk students: An open source analytics initiative. J. Learn. Analytics **1**, 6–47 (2014)

11. Corana, A., Marchesi, M., Martini, C., Ridella, S.: Minimizing multimodal functions of continuous variables with the "simulated annealing" algorithm. ACM Trans. Math. Softw. (TOMS) **13**, 262–280 (1987)

12. Cohen, J.: A power primer. Psychol. Bull. **112**, 155 (1992)

13. Schohn, G., Cohn, D.: Less is more: active learning with support vector machines. In: ICML, San Francisco, pp. 839–846 (2000)

14. Horvitz, E., Breese, J., Heckerman, D., Hovel, D., Rommelse, K.: The Lumiere project: bayesian user modeling for inferring the goals and needs of software users. In: Fourteenth Conference on Uncertainty in Artificial Intelligence, pp. 256–265. Morgan Kaufmann Publishers Inc. (1998)

15. Romero, C., Ventura, S., Espejo, P.G., Hervás, C.: Data mining algorithms to classify students. In: 1st International Conference on Educational Data Mining, Montreal, pp. 8–17 (2008)

16. Bäck, T., Schwefel, H.-P.: An overview of evolutionary algorithms for parameter optimization. Evol. Comput. **1**, 1–23 (1993)

17. Arnold, K.E., Pistilli, M.D.: Course signals at Purdue: using learning analytics to increase student success. In: 2nd International Conference on Learning Analytics and Knowledge, pp. 267–270. ACM, Vancouver (2012)

18. Krumm, A.E., Waddington, R.J., Teasley, S.D., Lonn, S.: A learning management system-based early warning system for academic advising in undergraduate engineering. In: Larusson, J.A., White, B. (eds.) Learning Analytics, pp. 103–119. Springer, New York (2014)

19. Gašević, D., Dawson, S., Rogers, T., Gasevic, D.: Learning analytics should not promote one size fits all: the effects of instructional conditions in predicting academic success. Internet High. Educ. **28**, 68–84 (2016)

20. Watson, I., Marir, F.: Case-based reasoning: a review. Knowl. Eng. Rev. **9**, 327–354 (1994)

Building a Mental Health Knowledge Model to Facilitate Decision Support

Bo Hu[✉] and Boris Villazon Terrazas

Fujitsu Laboratories of Europe, London, UK
{bo.hu,boris.villazon.terrazas}@uk.fujitsu.com

Abstract. Medical research produces a vast amount of data every-day through for instance high throughput preclinical and clinical tools. Exploiting such a source of knowledge, as well as discovering patterns and relations buried within, can offer great help to clinical profession-als in high quality health care services. There is a growing reliance on advanced computing technologies to help make sense and comprehend such data. In this paper, we describe the application of Word2Vec to facilitate knowledge discovery from very-large public unstructured text corpora (worked with PubMed thus far, but can easily incorporate oth-ers). Benefit from unsupervised word embedding, we experiment how new knowledge can stem from peer-reviewed medical publications and cross-reference such knowledge with established one to understand the advantages and disadvantages of popular deep-learning based approaches to knowledge acquisition. We also developed a proof-of-concept computer system to exploit such knowledge in a medical recommendation system.

1 Introduction

In order to provide high quality health care, it is vitally important for clinical professionals to "grasp" the latest developments in areas related to his/her exper-tise. Conventionally, this is achieved mainly through experience, self-training, and reflection, involving browsing through thousands of paper or electronic pub-lications. Documents in the scientific literature have long been serving as a source for knowledge transfer and knowledge discovery in terms of concepts (being for instance diseases, symptoms and treatments) and relationships among such con-cepts. Without the support of computers, however, exploring such a vast knowl-edge repository proves to be impractical. It is our contention that information overload and irrelevant information are major obstacles for drawing conclusions on personal health status and taking high-quality, adequate actions. The needs for ICT support fuel an increasing interest from both academic and industrial organisations to facilitate efficient access to information reported in scientific publications, patents, authority reports, and even from public portal web sites.

Medical publications and open data. Medical publication repositories such as MEDLINE currently index more than 22 million references [20]. MEDLINE is

© Springer International Publishing Switzerland 2016
H. Ohwada and K. Yoshida (Eds.): PKAW 2016, LNAI 9806, pp. 198–212, 2016.
DOI: 10.1007/978-3-319-42706-5_15

searchable through PubMed[1] (hereinafter, PubMed is used to refer to MEDLINE database for simplicity). Through the PubMed API, one could access the latest clinical research outcomes from academic and industrial organisations using natural language based queries. Automated information extraction applied on such unstructured text data can emerge patterns and relationships, which are not readily available to human readers. These can then be treated as scientific evidences to facilitate decision making. Applying data mining/text mining techniques to published literature has already been widely adopted in medical domains, e.g., for hypothesis generation [19], relation discovery [6], and pathway (re)construction [13]. Apart from PubMed, other open data sources, such as NICE[2] and openFDA[3] also played important roles in the development towards innovative healthcare solutions.

In general, we believe that PubMed is ideal for latest clinical research outcomes. This is particularly true for genetic and protein analysis. PubMed, however, maybe not be ideal for reflecting the current practice and established routines. In this paper, we report the initial results of exploiting a knowledge extraction tool built upon PubMed, which can be easily expanded to other data sources. This aims to help clinical professionals (as well as ordinary users) to deal with information overload and in some cases the so-called disinformation in the big data era.

Use case: mental health. Though the technology presented in this paper can be applied to any clinical areas, we choose mental health as the exemplar clinical domain. This is mainly due to the following theoretical and practical considerations. Mental disorder is a leading disease burden estimated by World Health Organisation (WHO) and yet given the same level of attention as many physical diseases [21]. In UK and major western European countries, mental illness is considered one of the biggest challenges of modern society. It is estimated that one in four residences in the UK is directly affected by mental illnesses while 27 % of the total adult EU population experienced a certain type of mental disorders. The impact of mental illness is not only manifested through decreased productivity and quality of life, but also reflected in a high rate of mortality. Mental health, however, is lagging behind other clinical areas in terms of technology support and budget [8]. Innovative ICT solutions, therefore, can significantly impinge upon the current practice. On the other hand, mental disorders present many unique characteristics that may render solutions borrowed from physical illnesses less effective. Meanwhile, in the past decade or so, research in areas such as genetics, behaviour/environments, and drug-induced problems has led to new discoveries and treatments [4,11]. It is essential to channel such latest breakthroughs into everyday patient care.

Symptoms and diseases. Despite the rapid advance of our knowledge about genetics and genomics, such new technologies have yet to become routine measures in everyday patient consultation due to the cost and availability. Clinicians

---

[1] http://www.ncbi.nlm.nih.gov/pubmed.

[2] http://www.nice.org.uk.

[3] http://open.fda.gov.

and general practitioners still reply heavily on symptoms and signs to help with their decision making. Symptom-disease connections, however, in many cases are not clearly defined. This is particularly troublesome in mental healthcare due to comorbidity of physical and mental disorders, difficulties in verbal communication, ambiguity of everyday language, and a lack of reliable, specific biomarkers. Previous efforts mainly focused on generic symptom-disease and disease-disease networks: either as manually curated ontologies [15] or through semi-supervised text mining [22]. Issues with these existing approaches lie in their dependence of domain knowledge and/or feature engineering. It is our contention that in order to effectively and efficiently integrate the best from both medicine and ICT, a fully automated system is desirable. Such a system should (i) substitute individual domain experts during the ingestion of the vast amount of data available today, (ii) minimise human intervention in the knowledge extraction process, and (iii) accommodate both human and machine users.

The remainder of this paper is organised as follows. In Sect. 2, we provide a brief overview of existing technologies for extracting knowledge from large unstructured text corpora. In Sect. 3, we explain details of data collection and model training using Word2vec and take a close look of the resultant word vector representations. In Sect. 4, we present a key use case of the word vectors in a clinical decision support system. We conclude the paper in Sect. 5 with potential improvements and key future directions.

2   Related Work

Text mining applies technologies such as natural language processing (NLP), information retrieval (IR), and data mining to discover patterns and trends from unstructured text. Text mining in biomedical domain has been extensively investigated [14]. A typical text mining pipeline starts with data collection; the raw data then go through text pre-processing, modelling (unstructured to structured data), and application/evaluation. Pre-processing of unstructured text largely dictates the overall quality of the outcomes at the end of the pipeline. Typical pre-processing tasks are tokenisation, stop words removal, and stemming. The modelling stage normally involves named entity recognition (NER) and relation extraction.

Thus far, in biomedicine, many tools have been developed to tackle each and every step of the text mining pipeline [7,9,10,18]. Conventionally, the underlying technology for lifting unstructured data to structured ones relies heavily on predefined, largely manually crafted *dictionaries* or ontologies (such as UMLS for medical applications) [2]. In 2011 and 2013, MAVIR research network and University Carlos III of Madrid, Spain organised two information extraction challenges targeting specifically on drug information extraction [16,17]. Both of the two challenges depended on labelled corpora for machine learning-based named entity recognition (NER). Similar endeavors exist in disease name recognition [3], underpinned by ontologies coming out of national or community-wide collaborations. Advantages and disadvantages of conventional approaches are equally evident. On the one hand, it ensures high quality outcomes based on carefully maintained, high quality data. On the other hand, the compilation of labelled corpora requires manual inputs from human data inspectors/curators. It is time consuming and can be problematic when it is either impossible or impractical to define a comprehensive *dictionary* – even if such global models can be negotiated, they are either too generic whereas adaption in specific sub-domains is necessary (*c.f.* the US modification ICD9-CM) or so colossal that one has to trim for real-life applications.

Moreover, medical domain in particular renders such reliance on predefined dictionary inefficient due to the depth and breadth of the domain of discourse, the rapid evolution of the domain expertise, the great diversity of clinical presentations, and the subtle yet crucial variations in human perceptions. In this paper, we investigate and implement tools that can acquire knowledge from a vast amount of non-confidential and scientifically-proven medical data gleaned from public domain. More importantly, we aim to facilitate such knowledge acquisition with only limited resources and minimal a priori domain knowledge, during the early development and roll-out cycles.

3   Learning a Mental Health Knowledge Base

The standard pipeline is illustrated in Fig. 1. Briefly speaking, data are collected from PubMed through the official API. The system then applies some basic techniques in the preprocessing module which is followed by Word2vec [12] based language model training. We experimented several sets of training parameters and evaluated the results with a set of manually composed test data. As we emphasise minimal input from domain experts, we draw domain knowledge from

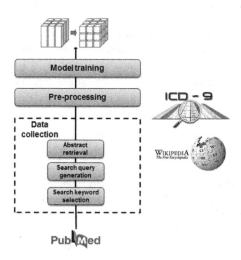

Fig. 1. Full pipeline

the public domain: both ICD-9 taxonomy and Wikipedia lists are used in data collection while the latter is also used in preprocessing.

3.1 Data Collection

We have experimented two data crawling strategies: indiscriminate data dumping and targeted data collection. In the first case, a public dump of PubMed was used (obtained through PMC FTP[4], containing articles from the PMC Open Access Subset). In the second case, only data concerning mental disorders were retrieved. Targeted data collection is carried out using pre-compiled lists of mental disorders. More specifically, in order to achieve full automation of the entire pipeline, we utilise international standards complemented with collaboratively created Wikipedia lists. Targeted data collection consists of three steps: search keyword selection, query generation, and abstract retrieval.

Search Keyword Selection. Key phrases used in exploring PubMed were collected as follows:

1. Listing 290-319 (mental related) from WHO International Classification of Diseases (ICD) version 9 were used as the primary guidance for generating search queries to PubMed web API. The first and second levels of headings were used after basic NLP processes (such as generic Stop word removal). The resultant set is denoted as S_{ICD}.
2. Items listed in Wikipedia under the `List_of_mental_disorders` (denoted as S_{W}) were used to compliment the ICD9 listings.
3. For each item in S_{W}, the web crawler followed the link and retrieved the title and the first paragraph of the actual Wikipedia document. The title was used to fine-tune the search keywords and the first paragraph was to enrich search keywords, for instance by adding to the key phrase set those introduced through "also known as [.*]+" or "also called [.*]+" patterns (denoted as S_{A}).
4. $\bigcup_{x \in \{\text{ICD, W, A}\}} S_x$ was cleaned up as follows:
   (a) we removed and/or replaced all non-alphabet characters;
   (b) for $w \in \bigcup_{x \in \{\text{ICD, W, A}\}} S_x$, for each w_j where $w_j \in \bigcup_{x \in \{\text{ICD, W, A}\}} S_x$ and $w \neq w_j$, we calculated $\Delta(w, w_j)$ as a real numeric value;
   (c) when distance between w and w_j, $\Delta(w, w_j)$ is sufficiently small, we removed w_j, to reduce the size of the search key phrases.

Note that the present solution is not ICD9 specific. ICD version 10 could have been used in the first step. Since the entire pipeline is fully automated, incorporation of further ICD versions as well as other taxonomies should be straightforward. MeSH taxonomy (subtree C10) could have guided the key phrase selection. We decided against using MeSH based on the observation that internally PubMed leverages MeSH in query handling. We use the ICD9 descriptions rather than codes due to availability: unfortunately, only a small number of clinical articles are annotated with ICD codes; searching PubMed directly with disease codes will not generate enough results.

---

[4] http://www.ncbi.nlm.nih.gov/pmc/tools/ftp/.

Search Query Generation. Search phrases are concatenated to compose PubMed queries. Internally, PubMed maps search phrases to MeSH ontology to exploit the semantics and conducts preliminary *query rewrite* so as to improve the search quality. The performance of official PubMed *entrez* web API can vary from time to time. In practice, it is possible to set the return size to 50 K and re-attempt when "404" or "502" error message occurs. An exemplar HTTP search query is as follows:

```
<pubmed web api>/esearch.fcgi?db=pubmed&retmode=json
                        &retmax=50000&term=<query phrase>
```

Abstract Retrieval. When carrying out keyword based search, PubMed matches the given search phrases to the title, keywords, and abstract of an article. More sophisticated search options can be applied to size down the return list. Keyword search only returns IDs of matching articles. PubMed provides another HTTP API to retrieve details (e.g., authors, publication date, abstract, etc.) of a given article in plain text or XML. We used fast redundancy check algorithm (e.g., Bloomfilter) to ensure duplicate article IDs are removed so as to improve system performance. For training the language model, only abstracts are pooled. The resultant text corpus is around 0.2 b words and 2.8 b characters.

3.2 Pre-processing

The retrieved text data are subject to several basic pre-processing steps: tokenisation, case folding, and basic plural stemming. Word2vec only embeds single words where disease names can consist of multiple words, e.g., "eating disorder". A key pre-processing step, therefore, is to concatenate disease names with hyphens, changing "eating disorder" into "eating-disorder". Full disease names are obtained from Wikipedia and ICD taxonomies. Due to the presence of name variations, it is also necessary to unify disease names before word embedding can be performed. Wikipedia in this case is used to unify names following the "x *also known as* x_n" and "x *also called* x_n" linguistic patterns. After pre-processing, we end up with around 60 different types of mental disorders.

Overall, the preprocessing step has reduced the vocabulary size of the text corpus from just above 400 k to around 230 k. Other typical text preprocessing steps are not exercised. For instance, stop word, punctuation removal and complicated stemming were not applied to the text corpus. The reason is two-fold. Firstly, we would like to reduce human intervention along the whole pipeline while tasks such as truncation currently involve considerable human supervision. Secondly, Word2vec exploits in-sentence word proximity information to understand the context of words during model training.

3.3 Training

Word embedding is carried out using Word2vec. Word2vec, open-sourced by Google, is a two-layer (one visible and one hidden layer) neural language model that encodes words as high dimensional vectors. Word2vec takes as input large datasets of unstructured text data, ignoring the grammar, and tries to compute continuous distributed vector representations of words, effectively projecting the words onto an n-dimensional space (where n is normally greater than 100). These word vectors stem from the contexts of words, i.e., a number of words appearing before and after the target word. Word context is utilised through two mechanisms: *continuous bag of words* (cbow) and *skip-gram* (sg) – in cbow, influence of context words is used to update the representation of the target words while in the latter, target words impinge on the context words. Based on the literature and also confirmed with our experiment, *skip-gram* produces better vectors for infrequent words, making it preferable in working with medications and symptoms. Word vector models produced by Word2vec are not universal. The values of vector representations (and thus their projections upon the vector space) highly depend on the training data from where the vectors are calculated and only reflect the distribution and word associations within the context of the training text corpora. During training, the pre-processed text corpus is fed into Word2vec to (i) extract a vocabulary from the training text data, (ii) obtain and refine the language model, and (iii) for each word, produce a vector of real numbers. Word vectors can then be used as features in many NLP and machine learning applications. One typical application is to quantify the distance of two words using Cosine measures. Meanwhile, a word vector model also facilitates data clustering. For instance, a K-means clustering algorithm can be easily implemented exploiting the Cosine distances of different word vectors.

We have experimented with different context windows and vector sizes. The quality of outcomes with different parameters is evaluated using a set of manually labelled data (denoted as Θ_{eval}), which is obtained as follows: (i) twenty common mental disorders are randomly selected (including *parkinson*, *alzheimer*, *autism*, etc.) and denoted as θ_D; (ii) for each disease, the top 10 symptoms are compiled from guidelines, textbooks and domain expert advices; (iii) also for each disease, the top 5 currently most widely used medicines are collected. The list of medicines contains only generic names and is composed based on NICE recommendations, FDA guidelines, and Wikipedia.

Precision and recall are the criteria for judging the quality of outcomes. Precision and recall are defined in a slightly different sense from standard ones: *precision* – how many in the top 50 results that are computed based on Word2vec model are correct; *recall* – how many of the manually compile data have been included in the top 50 results. This deviation (formalised in Eq. 1) from standard term usage is based on both theoretical and practical considerations. On the one hand, computing word vector is to acquire significant relations among all the words appearing in the raw text. It, therefore, has different a rationale from information retrieval and pattern recognition tasks wherein the two criteria originated. On the other hand, we resort to these measures for system tuning instead of full performance evaluation. Simply applying distance measures on top of the resultant vectors can bring in noise data and hence make the conventional precision and recall measure less meaningful. Non-stringent, relative measurement is more appropriate. Note that Θ_{eval} is by no means to provide a comprehensive evaluation data sets to fully interrogate the performance of Word2vec. Instead, it provides a means to quickly gauge and optimise the parameter settings.

$$\text{precision} = \frac{\Theta_{\text{top50}} \cap \Theta_{\text{eval}}}{\Theta_{\text{top50}}}, \quad \text{recall} = \frac{\Theta_{\text{top50}} \cap \Theta_{\text{eval}}}{\Theta_{\text{eval}}}$$

$$\Theta_{\text{top50}} = \bigcup_{\forall x \in \theta_D} \left\{ y_i | y_i \in \{\cos(x, y_n)\}_{n=1}^{50} \right\} \tag{1}$$

where θ_D is the set of selected diseases, Θ_{eval} evaluation dataset, Θ_{top50} the top 50 results calculated based on word vector models, and $cos(x, y)$ computes cosine similarity of x and y.

3.4   Results and Discussions

We are interested in the relationships among diseases, their indicating or dominant symptoms/signs, and their medications. One of the key advantages of a continuous real-valued vector representation is the possibility of applying vector algebra to compute for instance distances and similarities among words.

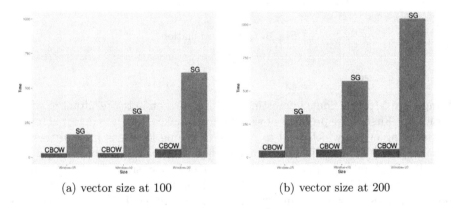

(a) vector size at 100   (b) vector size at 200

Fig. 2. Time performance

Tuning Parameters. Several model configurations are tested (including different values of dimensionality and context window size). The two context incorporating mechanisms (i.e., cbow and sg) impinge both performance and outcomes. Figure 2 illustrates time consumption of the two mechanisms with different sizes of context window (5, 10, and 20 respectively) and different sizes of resultant word vectors (100 and 200 respectively). It demonstrates that cbow runs 2 to 3 times faster than sg with the same iteration and negative sampling settings. The hardware configuration is: Fujitsu PRiMERGY RX200 S6 server with 2 quad-core Intel CPUs at 2.40 GHz.

When evaluated against Θ_{eval}, it is evident that sg in general presents higher precision and recall values comparing with cbow while different sg configurations produce outcomes of similar quality (shown in Fig. 3, where dark coloured bar shows precision and light coloured recall). Taking also into account the time factor, we conclude that the combination of skip-gram, 10-word window size, and 200 vector size seems to be the best setup.

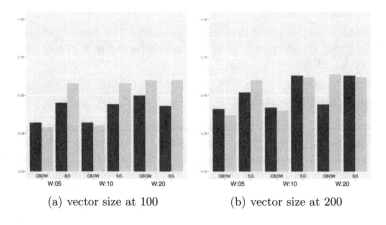

(a) vector size at 100   (b) vector size at 200

Fig. 3. Quality evaluation

Language Model. Some interesting findings are seen in the resultant word vector model. Firstly, the presence of word context makes it easy to detect acronyms and spelling errors which tend to occupy the same locations in sentences. For instance, both "huntingtin" and "huntinton" are considered to be strongly associated with "huntington" disease. AD and HD are considered the top matches of Alzheimer and Huntington where Schizophrenia is frequently short-handed as both "SZ" (cosine = 0.848) and "SCZ" (cosine = 0.801). This may appear to be trivia. However, the vectors are generated without a priori domain knowledge and can be easily applied to other domains and used by users with only limited domain knowledge.

The simplest model utilisation is to compute how closely two diseases or two symptoms are related. This is tantamount to compute the cosine similarity between two word vectors in the space defined using PubMed text corpora. For instance, alzheimer and huntington's diseases are considered highly related ones of parkinson's diseases. Meanwhile, querying the model with *tremor* produces the following top five clinical terms each with relatively high similarity: bradykinesia (0.773), akinesia (0.694), dystonia (0.646), hypokinesia (0.641), and akinetic-rigidity (0.637).

Symptom and Disease. Unlike physical illnesses, mental disorders present only a limited number of identifiable causative agents and observable alteration of physical organs. A symptom, therefore, can be demonstrated in multiple diseases with different extend of magnitude and significance of indicating values. For instance, "tremor" presents in a whole raft of mental and neurological diseases. The word vector can be used to find diseases that are strongly associated with a symptom. As shown in Table 1, if one knows tremor is a leading symptom of Parkinson's disease, other symptoms can be easily computed based on the distance of the corresponding vectors: $\mathbf{V}_{tremor} - \mathbf{V}_{parkinson}$ and $\mathbf{V}_{tremor} - \mathbf{V}_i$ where $i \in \mathbb{D}$, $\mathbb{D}$ is the set of diseases obtained from public data (as discussed in previous sections). As illustrated in Table 1 left columns, leading diseases are filtered out with the assistance of a mental disease dictionary.

Table 1. Diseases vs. Symptoms

Tremor		Parkinson		
Diseases	cosine	Symptoms	"symptom"	"tremor"
parkinson	-	jittery	0.635	0.635
huntington's disease	0.674	dizziness	0.605	-
alzheimer	0.648	tremble	0.599	0.599
SWEDD	0.630	tachypnea	0.598	-
epilepsy	0.532	distress	0.590	-
dementia	0.531	numb	0.589	-
pain disorder	0.483	palpitation	0.587	-
insomnia	0.481	anxiety	0.585	-

Finding symptoms for diseases, however, is not so straightforward. The following two strategies are experimented: representative and anchor word. A typical symptom is selected against which cosine values are computed. In the case of Parkinson's disease, "tremor" is selected and cosine values of $\mathbf{V}_{tremor} - \mathbf{V}_{parkinson}$ and $\mathbf{V}_x - \mathbf{V}_{parkinson}$ are computed. By examining the results, this approach seems to present two problems: firstly, a predefined set of words from where x is drawn should be defined to avoid screening all the vector space for candidates; secondly, one has to have some a priori knowledge of the domain to name the typical symptoms. The first problem can be partially solved by first identifying a subset of words that have very small distances $(1 - cos(V_1, V_2))$ from "tremor". This subset is then considered as candidate words for computation. The second problem, however, is against our intention to have minimal intervention from human domain experts and is prone to inter-expert variation.

The second strategy is to use *anchor* word such as "symptom" and treat $\mathbf{V}_{symptom} - \mathbf{V}_{parkinson}$ as the baseline regularity against which other words can be filtered. In practice, combine word clustering (as explained above) and this simple strategy using anchor word "symptom" seems to offer the strongest selective power (right columns of Table 1). Intuitively, "jittery", "dizziness", and "numb" refers to perceivable signs and observable symptoms; "distress" and "anxiety" on the other hand are boundary words for both symptoms and diseases. Others not shown in the table include malaise, paresthesia, talkative, perspiration, etc. some of which are not intuitively considered as typical signs/symptoms of parkinson.

Drugs. On the contrary, the representative word strategy works better to identify medicines for diseases than the anchor strategy does. The vector space model becomes useful, when one has some knowledge about a disease and tries to draw analogy to another disease. One possible explanation of such a discrepancy is that authors may frequently use word "symptom" as the hint to enumerate symptoms, for instance "typical symptoms are ...", while the same linguistic patterns may not be widely adopted in scientific writing when discussing drugs. Also, multiple words can be used to refer to "drugs" (e.g., "medication", "treatment", etc.) making the detection even more difficult.

Anchor word based vector algebra: $\mathbf{V}_{\text{levodopa}} - \mathbf{V}_{\text{parkinson}} + \mathbf{V}_x$, where x is the disease name, helps to correctly identify "benzodiazepine" in the top ten words for anxiety and RLAI (Risperidone long-acting injection) for schizophrenia. Some of the disease-drug relation results are listed in Table 2.

Table 2. Diseases vs. Drugs

Alzheimer		Schizophrenia	
Drug	similarity	Drug	similarity
l-dopa	0.580	clozapine	0.593
d-galactos	0.575	blonanserin	0.562
leucovorin	0.568	RLAI	0.539
perindopril	0.566	olanzapine	0.526
entacapone	0.552	aripiprazole	0.521
carbidopa	0.551	risperidone	0.514
sinemet	0.546	haloperidol	0.509
benserazide	0.545	ziprasidon	0.507

Some points worth further discussions with respect to this strategy. Firstly, consulting a complete list of known drugs (with their brand and generic names) and compounds may improve the accuracy. For such a purpose, several open and commercial communities maintain up-to-date lists of generic names of medicines, e.g., SNOMED CT, the Drug Ontology [5], FDA API, Merck Manuals [1], etc. This is particularly useful when the names consisting of multiple words. Secondly, unlike physical diseases, mental disorders frequently resort to non-drug therapies, e.g., cognitive behaviour therapy, which can negatively affect the efficiency of the above approach. Exact impact is yet to be quantified (as our immediate future work). Thirdly, publications available through PubMed focus primarily on new and exploratory methods instead of established treatments. Solely relaying on PubMed-based text mining can bias the results towards non-conventional, innovative, or even "disastrous" treatments. In the decision support system, therefore, the results are used primarily as non mission-critical evidences for users to explore alternatives.

4   Consuming Word Vector Model

The goal of acquiring a language model is to estimate the probability word association so to serve as evidence in higher level decision making. In this section, we present a system, utilising similarity and distance measures of word vectors, to assist in clinical decision making and to enhance the training and education of clinical professionals.

4.1   Architecture

The decision support system is underpinned by the Word2vec-based language model obtained as in previous sections. Fusing also knowledge from other sources, the system tries to highlight evidences that can support clinical decision making. The overall system architecture is shown in Fig. 4. Core components of the system include: data source management, model generation, data storage, and an end user dashboard.

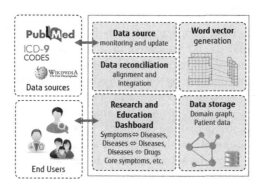

Fig. 4. System architecture

Data source management is to monitor the status of selected public data sources (being PubMed and ICD classification at the moment) and periodically pool data from such sources to update local data store. Some basic data reconciliation (e.g., string-based and taxonomy-based matching techniques) and preprocessing techniques are performed to reduce ambiguity of the data and to align entities residing in different domains. Should the external data sources change, the word vector model will be reconstructed. Re-training is performed in two cycles. Minor model updates will be carried out at the end of each data pooling cycle with the optimal parameters discussed in the previous sections. Major model updates will be scheduled after several minor updates to re-evaluate all the parameters (using test dataset Θ_{eval}). The system only undergoes major updates when resource permits. In practice, major updates are done every three months.

The resultant word vector model is stored in its native binary format and exploited in two different manners. Firstly, it can serve as an "online" knowledge base, to which queries drawn from the anonymised patient data are posed. Typical scenarios include (i) recommending medications for a given diagnosis, (ii) for one observed symptom, suggesting other related ones to examine, and (iii) for one diagnosis, highlighting potential comorbid conditions. Such an online process is interactive based on users' inputs. Secondly, the word vector model can be used in an "off-line" fashion to construct and refine a domain knowledge model obtained through other channels. For instance, with established ontologies and domain graph models (such as SNOMED CT, where entities are represented as graph vertices and relationships as graph edges), word vector model can be used to confirm or fine-tune relationships defined in other models. For an existing relationship $\langle u, v \rangle$, word vector model can quantify the edge strength by computing the similarity of the vectors corresponding to vertices u and v. This can easily transform the original model into a weighted graph, giving rise to a whole raft of graph-based inferences.

4.2 Evaluation Plan

Evaluation of the decision support system is forthcoming. Key features of the system will be evaluated and validated in a workshop with selected domain experts and other types of end users. The workshop will consist of a walk-through and hands-on session followed by on-site feedback and off-site questionnaires to solicit comments on system design and usability. If a positive outcome is achieved, the system can be deployed at clinical wards as a stand alone tool. As we target only at secondary use of clinical data, the system is not expected to undergo extensive clinical trial, nor a full integration with current hospital information systems. It can be set alongside other educational and research tools that are currently in use at hospitals.

5 Discussions and Conclusions

Biomedical knowledge grows exponentially in both depth and breadth. In the past years, it has been revealed that each year around 700 K new journal articles are added to MEDLINE, the main medical literature database. There are yet thousands of other types of publications, e.g., clinical reports, conference posters, authority guidelines, made available to public each month. Without the help of computer systems, it is hard for such invaluable data to find their way into the everyday patient care. In this paper, we propose a system to (i) extract and encode the knowledge buried in medical publications using neural network language models and (ii) leverage such a language model in a decision support system. Our preliminary evaluation confirms the viability of a fully automated, unsupervised text mining and knowledge acquisition approach. The word vector model is able to reveal key relationships between symptoms and diseases. Meanwhile, with clustering and potentially the help of dictionaries (for drugs and proteins, for instance), it can also draw connections between diseases, medicines/treatments, and target proteins. The word vectors can be exploited to deliver the latest research outcomes as secondary evidences for clinical decision making.

The proposed approach is subject to more comprehensive evaluation. We also expect to further develop along the following improvements and enhancements.

Full evaluation: Thus far only preliminary evaluation has been carried out against the full pipeline technology. The crux of our immediate future work is soliciting domain experts to perform thorough evaluation of both the quality of word embedding and the usability of the knowledge-based system. Initially, dedicated hand-on workshop sessions are to be organised with experienced clinical professionals; if positive, roll-out to production environment will be considered.

Other data sources: Though some interesting findings with respect to symptoms and signs have been observed with word embedding based on PubMed, the coverage is not satisfactory. This is partially due to the fact that many MEDLINE research articles focus more on pathology, epidemiology, pharmacology of diseases and medications, instead of surveying full set of symptoms and signs. In order to achieve better results, it is necessary to train the word vector model with data that contain fair presences of all types of disease symptoms and signs. A promising source could be patient-centric portal web sites, such as *patientlikeme.com, patient.info*, etc. and authority information sites such as NHS patient portal site.

Other domains: In this paper, we focus on mental health to demonstrate the concept and applicability of acquiring knowledge from seemingly untamable big unstructured data. In order to work with other clinical domains, the word vector model will have to be retrained with appropriate datasets. We, nevertheless, believe the fundamental idea can be easily implemented in other clinical and non-clinical areas.

References

1. The Merck Manuals: of diagnosis and therapy. Merck Sharp & Dohme, Whitehouse Station, N.J. (2009)
2. Danger, R., Segura-Bedmar, I., Martinez, P., Rosso, P.: A comparison of machine learning techniques for detection of drug target articles. J. Biomed. Inform. **43**(6), 902–913 (2010)
3. Dogan, R.I., Leaman, R., Lu, Z.: NCBI disease corpus: a resource for disease name recognition and concept normalization. J. Biomed. Inform. **47**, 1–10 (2014)

4. Doherty, J.L., Owen, M.J.: Genomic insights into the overlap between psychiatric disorders: implications for research and clinical practice. Genome Med. **6**(4), 29 (2014)
5. Hanna, J., Joseph, E., Brochhausen, M., Hogan, W.R.: Building a drug ontology based on RxNorm and other sources. J. Biomed. Semant. **4**, 44 (2013)
6. Jaeger, S., Gaudan, S., Leser, U., Rebholz-Schuhmann, D.: Integrating protein-protein interactions and text mining for protein function prediction. BMC Bioinform. **9**(8), 1–10 (2008)
7. Jiang, J., Zhai, C.: An empirical study of tokenization strategies for biomedical information retrieval. Inf. Retr. **10**(4–5), 341–363 (2007)
8. Kilbourne, A.M., Keyser, D., Pincus, H.A.: Challenges and opportunities in measuring the quality of mental health care. Can. J. Psychiatry **55**(9), 549–557 (2010)
9. Leaman, R., Gonzalez, G.: Banner: an executable survey of advances in biomedical named entity recognition. In: Altman, R.B., Dunker, A.K., Hunter, L., Murray, T., Klein, T.E. (eds.) Pacific Symposium on Biocomputing, pp. 652–663 (2008)
10. Liu, H., Christiansen, T., Baumgartner, W.A., Verspoor, K.: BioLemmatizer: a lemmatization tool for morphological processing of biomedical text. J. Biomed. Semant. **3**(1), 3 (2012)
11. Merikangas, K.R., Risch, N.: Will the genomics revolution revolutionize psychiatry? Am. J. Psychiatry **160**(4), 625–635 (2003)
12. Mikolov, T., Sutskever, I., Chen, K., Corrado, G.S., Dean, J.: Distributed representations of words and phrases and their compositionality. Adv. Neural Inf. Process. Syst. **26**, 3111–3119 (2013)
13. Miwa, M., Ohta, T., Rak, R., Rowley, A., Kell, D.B., Pyysalo, S., Ananiadou, S.: A method for integrating and ranking the evidence for biochemical pathways by mining reactions from text. Bioinformatics **29**(13), i44–i52 (2013)
14. Rodriguez-Esteban, R.: Biomedical text mining and its applications. PLoS Comput. Biol. **5**(12), e1000597 (2009)
15. Schriml, L.M.M., Arze, C., Nadendla, S., Chang, Y., Mazaitis, M., Felix, V., Feng, G., Kibbe, W.A.A.: Disease Ontology: a backbone for disease semantic integration. Nucleic Acids Res. **40**(Database issue), D940–D946 (2012)
16. Segura-Bedmar, I., Martínez, P., Herrero Zazo, M.: Semeval-2013 task 9: extraction of drug-drug interactions from biomedical texts (ddiextraction2013). In: Second Joint Conference on Lexical and Computational Semantics (*SEM), Proceedings of the Seventh International Workshop on Semantic Evaluation (SemEval 2013), vol. 2, pp. 341–350 (2013)
17. Segura-Bedmar, I., Martinez, P., Sanchez-Cisneros, D.: Extraction of drug-drug interactions from biomedical texts. In: Proceedings of the 1st Challenge Task on DDIExtraction, pp. 1–9 (2011)
18. Settles, B.: Biomedical named entity recognition using conditional random fields and rich feature sets. In: Proceedings of JNLPBA 2004, pp. 104–107. ACL (2004)
19. Spangler, S., Wilkins, A.D., et al.: Automated hypothesis generation based on mining scientific literature. In: Proceedings of the 20th ACM SIGKDD International Conference on Knowledge Discovery and Data Mining, KDD 2014, pp. 1877–1886. ACM (2014)
20. U.S. National Library of Medicine: MEDLINE fact sheet. https://www.nlm.nih.gov/pubs/factsheets/medline.html (accessed in Jan 2016)
21. World Health Organization: The global burden of disease: 2004 update (2008)
22. Zhou, X., Menche, J., Barabási, A.L., Sharma, A.: Human symptoms-disease network. Nat. Commun. **5** (2014)

Building a Working Alliance with a Knowledge Based System Through an Embodied Conversational Agent

Deborah Richards[1(✉)] and Patrina Caldwell[2]

[1] Computing Department, Macquarie University, Sydney, NSW 2109, Australia
deborah.richards@mq.edu.au
[2] Centre for Kidney Research & Discipline of Paediatrics and Child Health,
University of Sydney, Sydney, Australia
patrina.caldwell@health.nsw.gov.au

Abstract. Knowledge is only useful if the intended user is able to utilize the system. In early knowledge based systems, known as expert systems, the user interface provided the means through which knowledge was acquired and accessed. Today it is possible to interact with the knowledge of an expert through a humanlike interface, in the form of an embodied conversational agent (ECA). Through familiar conversational-style interaction, the ECA can obtain the state of the user and provide a recommendation overcoming health literacy barriers and, depending on the nature of the dialogue, build a working alliance with the human that will encourage adherence to the advice. In this paper we describe the eADVICE system in the domain of paediatric incontinence that aims to improve adherence through the use of an ECA as the interface to the domain knowledge. Results of an initial pilot are provided showing that those who used the ECA achieved improved health outcomes and found the experience positive.

Keywords: Knowledge based system · Embodied conversational agents · eADVICE

1 Introduction

Early expert systems offered a means of capturing the knowledge of domain experts. Many of those systems, like MYCIN [1], concerned medical expertise. MYCIN was developed in LISP the early 1970s to identify which bacteria was causing an infection and then prescribe the appropriate course of antibiotics. This system was never used in practice, but it was found to provide accurate conclusions more often than disease experts.

An expert system is comprised of a user interface, in addition to the knowledge base and inference engine. The importance of getting the user interface right was already evident in early systems such as ONCOCIN, for managing the treatment of oncology patients, where it was found that telling a specialist what to do was not acceptable, however providing a critiquing style interface that provided a second opinion on the specialist's recommendation was appropriate [2].

© Springer International Publishing Switzerland 2016
H. Ohwada and K. Yoshida (Eds.): PKAW 2016, LNAI 9806, pp. 213–227, 2016.
DOI: 10.1007/978-3-319-42706-5_16

Four decades on, expert system technology is embedded in almost all artificially intelligent software, often in the form of one or more knowledge-based systems (KBS). KBS are no longer the main focus of the system but a component. However, interacting with that knowledge in a usable and useful ways still poses challenges.

As in the case of ONCOCIN a key issue is understanding how the user, whether they are a medical professional or a patient, will best receive and act on the knowledge provided. This problem is very evident when we see that reported adherence rates to treatment advice is around 50 % [3]. There is evidence to suggest that treatment adherence can be improved with face-to-face communication between a patient and health professional [4, 5] and particularly when a therapeutic alliance is built [6]. Given that health care resources to support face-to-face communication is extremely limited, we explore the efficacy of a user interface involving a humanlike embodied conversational agent (ECA). ECAs, particularly those who exhibit empathic behaviours, can potentially deliver non-judgmental support similar to real face-to-face communication [7, 8]. Evidence has shown that for some conditions, empathic agents are even more successful in communicating and encouraging treatment adherence [5, 9], thought further research may be necessary. However, the use of empathic agents for paediatric healthcare has been largely unexplored.

This paper presents the results of a pilot involving real paediatric patients and their families that used an empathic ECA to improve adherence to personalised treatment advice provided by an interactive website. The project is introduced in the following section followed by a review of the literature that informed design of the character's appearance and behaviours. A pilot study is presented in Sect. 3, followed by discussion, conclusion and future work in Sects. 4 and 5, respectively.

2   eADVICE

The Children's Hospital Westmead (CHW) has been trialling a web application named eADVICE (electronic Advice and Diagnosis Via the Internet following Computerised Evaluation) that prescribes treatment plans to child patients with urinary incontinence. Urinary incontinence was selected as an ideal condition for testing, as bed wetting is a common condition among children aged 5 to 18 years and is also relatively benign. Urinary incontinence occurs in 20 % of school aged children, and up to 2 % of adults [10]. Following years of medical practice, the specialist has identified seven possible treatments and the factors that determine when each treatment is appropriate. Further, she has formulated algorithms that determine the values of each factor and a set of rules associated with each treatment and action plan. Figure 1 depicts the 7 rulesets, one for each possible treatment, that form the eADVICE knowledge base. The Caffeine Advice; Fluid Advice; Bowel Program; and Timed Voiding rulesets have 2–5 rules in them. The rules associated with Enuresis Alarm Training and Medication KBSs are more complex and the conditions in some rules rely on the results of the patient following other treatments. The data that is collected can be shared with a patients' general practitioner to assist with decision making while the patient is awaiting a specialist appointment. Patients are intended to use eADVICE on a regular basis (e.g. every two weeks) until they become dry.

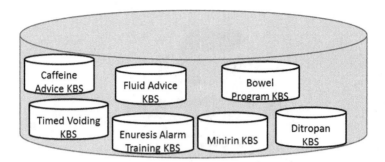

Fig. 1. eADVICE knowledge base coverage and rulesets.

Adherence to the recommendations provided by eADVICE was around 50 %. An empathic agent was added as the interface for the web application with the aim to improve adherence to treatment plans, build a working alliance with the patient and overcome health literacy boundaries. The empathic agent seeks to provide more timely and accessible treatment advice while also addressing the problem of lack of adherence.

3   Design of the Empathic Agent

The design of the ECA draws on themes emerging from a literature review to understand adherence: working alliance, verbal and non-verbal communication; formulating short term goals, empowering the patient, overcoming health literacy barriers, personalized treatment advice, monitoring of the patient's adherence.

A distinguishing characteristic of empathic agents in comparison to other self-help e-Health interventions is the establishment of the agent as an empathic, mentor-like figure and care partner for the patient. The reason for developing a relationship with the patient is to build trust and rapport with the agent so that the patient is more likely to be accepting of goals established with the agent [11]. Beun [11] highly recommends that in order to establish a socio-emotional association between the patient and agent, there should be a stage in the initial consultation or at least the early stages of therapy where the agent explicitly acknowledges the patient so that they can learn about one another.

We created a dialogue flowchart, presented in [12] to help us incorporate a number of empathic cues [13] into each of the dialogues associated with the seven treatment plans. A sample flowchart is shown in Fig. 2. These cues included: (a) empathy for the user (b) social dialogue (c) reciprocal self-disclosure (d) humour (e) meta-relational communication (talk about the relationship) (f) expressing happiness to see the user (g) talking about the past and future together (h) continuity behaviours (appropriate greetings and farewells and talk about the time spent apart) (i) reference to mutual knowledge (j) inclusive pronouns, politeness strategies, greeting and farewell rituals [13, p. 7].

Verbal communication strategies such as these can be effective techniques for creating awareness about their behaviour and realizing the discrepancy between their current actions and their desired goals [14]. The dialogue appears as speech bubbles, with options for selection for interaction. We chose not to use TextToSpeech (TTS) to

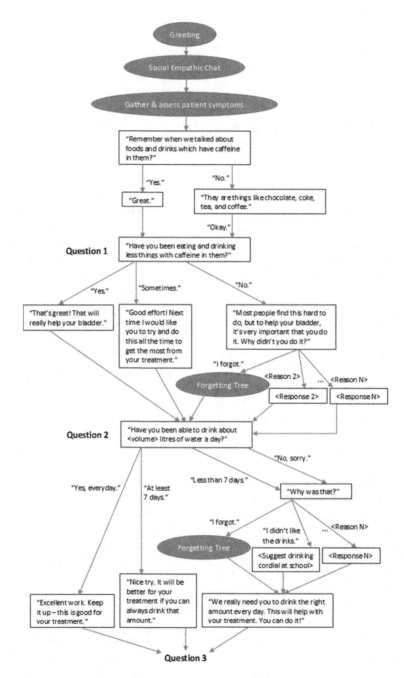

Fig. 2. Questions that the agent asks on the second or subsequent visits to check adherence.

generate spoken dialogue because people can strongly dislike automated speech and we thought it might harm the flow and sense of immersion that we thought was important for development of the therapeutic alliance. Instead sound bites using the specialist's voice have been recorded. As an added benefit, if patients did proceed with their specialist appointment, when their turn came, they would already be familiar with the doctor's voice.

Non-verbal communication is important to patients for clarifying the intent of spoken dialogue and that without it, phrases may be misunderstood or misinterpreted [15]. Non-verbal communication techniques include facial expression, hand gestures, posture, subtle head movements and other natural body language characteristics that would also apply to a human physician. To-date Dr Evie displays limited non-verbal behaviours consisting of lip synching and conveying a happy/positive demeanour on her face.

Dr Evie's dialogue recognises the importance of formulating short-term goals towards long term lifestyle changes. Goals are wants, desires, obligations or intentions of a patient to achieve a particular state [11]. These should be mutual goals between the patient and the agent, which can add to the development of a therapeutic alliance. Discussion can include identification of goals and appropriate rewards. A simple example is asking a patient suffering from incontinence to drink a specific volume of fluid for a given day. If they successfully follow the treatment, they could be rewarded by their parents for their positive actions.

Related to the empathic dialogue and development of a therapeutic/working alliance, we chose a female character. Female physicians have been found to be more positive and focus on emotions and partnership building [16, p. 17]. As a highly multicultural country, we also wanted our character to not be identifiable with any race or ethnicity, so we chose features less specific to any part of the world such as brown hair and brown eyes. The model went through many rounds of revisions with the medical and technical team. Figure 3 shows the final version of Dr Evie.

Fig. 3. Sample interaction with Dr Evie. The patient can stop at any time (top left) and see their progress (top right).

Perhaps the most central motif among the design of all empathic agents is the ability to motivate and empower patients. Traditional face-to-face appointments are limited by time and logistic constraints, among other reasons, which may significantly dampen a patient's motivation with long periods between consultations [17]. Maintaining a consistently high level of motivation is essential for patients for continued adherence to treatment advice [17]. While patients are not able to update their patient history or request another treatment plan until 2 weeks since the last plan, patients are able to interact with Dr Evie as often as they wish to discuss the treatment plan to gain further understanding of the reasons for the recommendations and to explore options. This makes Dr Evie more accessible than a specialist and allows clarification and discussion without fear of being found annoying or stupid.

Furthermore, an empathic agent can be used by a patient at any time that suits them, although it was noted by Cote et al. [18], patients require time to adapt and integrate change into their daily routine. Cote et al. therefore suggested that a minimum of 10 days between consultations with an agent would be enforced for their VIH-TAVIE intervention [18] to allow patients to adjust to lifestyle changes gradually. In our case, eADVICE was set up for interaction fortnightly so that patients would not return before giving the treatment a serious go.

Low health literacy is a barrier for accessing and receiving health information [19]. Increased exposure to empathic agents has been linked with patients' improved knowledge of their condition [20]. If a patient is unable to understand their treatment options, they are less likely to comply with the treatment, whether it is a conscious decision or accidental. Informing the patient and eliminating negative perspectives about treatment can inspire self-confidence to persist with the treatment.

In order to increase the likelihood of treatment adherence, the treatment plans must be tailored to individual needs of the patient [21]. Traditional self-care e-Health interventions offer generalised medical advice to patients based on previously known symptoms of a given condition that are targeted at broad audiences. This approach, however, does not account for the individual needs, abilities or knowledge of the patient [11]. Dr Evie only suggests treatments relevant to the patient history and also takes into account whether the patient has been given that advice before, whether they have been following the advice and if not, their reasons. Personalisation of treatment advice also implies negotiation between patient and agent. The dialogues allow patients to respond to the recommended advice to indicate their willingness to follow the treatment plan. Dr Evie will suggest alternative options or things for consideration.

Dreyer [22] noted that a significant advantage of an agent over traditional face-to-face consultations is that all formal processes that would normally be distracting to a patient (such as recording of medical history, wait times for doctors and administrative activities) can be integrated into the agent interaction. With a conversational character, natural language can be used to collect important patient data while under the guise of a normal conversation [17]. For example, it may be more engaging for an agent to learn about a patient's sleeping habits through dialogue rather than filling in text fields on a web page.

Families meet Dr Evie at the start and end of their consultation. As depicted in Fig. 4, eADVICE takes the medical history of the patient via web-forms plus time and

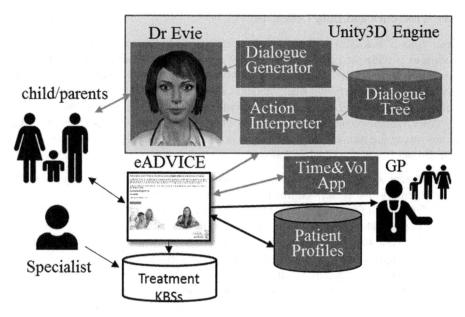

Fig. 4. Treatment assessment flow and system architecture. Black indicates original system and blue/grey indicates added components. (Color figure online)

volume data of fluid inputs and outputs to create a profile of the patient that is stored in a MySQL database. eADVICE uses an algorithm that encodes the expertise of the specialist to provide one or more of the seven possible recommended treatments based on the patient profile. To accompany each treatment, the medical team created dialogue trees taking into account their years of experience and the empathic queues [13]. The Action Interpreter and Dialogue Generator use the dialogue tree to drive Dr Evie via the Unity3D Game Engine. For lip synching a plugin to Unity3D has been used. eADVICE sends a report to the GP of the advised treatment.

4 Pilot Trial

It was our intention that all interaction would be via Dr Evie. However, the eADVICE system and associated diagnosis algorithms have been developed and refined over a period of many years so we did not want to reimplement the algorithms for the purpose of a pilot with real patients. eADVICE provides a link to the dialogue tree corresponding to each treatment. A sample of fourteen pediatric patients were recruited from the CHW waiting list and selected for the study. The selection criteria for the study included school aged children ranging between 5 and 18 years of age. Additionally, only English speaking families who owned a computer at home were selected for the study.

4.1 Procedure and Data Collection

Once consent has been obtained, patient and GP contact data are entered into the system to provide the patient with a userID and password. After logging in, Dr Evie introduces herself to let patients know that the treatment assessments can occur as frequently as once every two weeks and that after they have done the assessment in the online system, she will chat with them about their recommended treatment/s.

After a patient begins the assessment by clicking "Start Assessment", they are taken to the survey questions. This is the main component of eADVICE, which consists of five distinct stages described as follows:

1. Treatment Adherence Questions 1 | a set of questions that ask of the patient's adherence to the prescribed treatment plans from the previous session.
2. Patient Health Outcome Questions 1 | questions to asked if the patient's condition has improved and if they have become dry. If the patient has become dry, there is no need to continue with further questions and the assessment ends. 1 only asked from the second treatment assessment and each subsequent assessment.
3. Time-and-Volume Chart 2 | the patient is asked to complete time-and-volume charts for at most two days between the previous and current session, with the option of skipping to continue with the assessment. The time-and-volume chart provides data about the volume of liquid ingested and the volume of urine passed for each hour of the day. This information is used to prescribe medication to the patient. NOTE: since the pilot we have implemented a website and mobile App for capturing this data, so that this step can be replaced by an upload from the website or App.
4. Patient History Questions | a set of thirteen questions that ask about the patient's past history, which contribute to the treatment plan prescribed. Two sets of phrasings were used: a "Parent" phrasing and "Kid's Mode" phrasing.
5. Show Treatment Plan | the final screen which prescribes up to seven different treatment options. These treatment options are based of the responses given by the patient for the time-and-volume chart/s and the patient history questions. It is also noted that at any point in the assessment, the text can be switched between "Parent's Mode" and "Kid's Mode", where "Kid's Mode" asked the same questions but with text suitable for a child patient.

With the addition of Dr Evie, along with each treatment plan an icon with Dr Evie's face can be clicked to engage in a conversation regarding the recommended treatment (see Fig. 5). Clicking this link is optional and not required by the patient.

A separate database collects statistics about the exact responses to the empathic agent for each session and the corresponding treatment plan. This information allows us to determine the dialogue path taken by each patient. This data will be used to identify what responses are most common and identify patterns and relationships between their adherence and behavioural responses and the patient history.

In addition to collecting patient responses directly through eADVICE, responses to several sets of survey questions are gathered as pre and post study data collection. The feedback from the surveys is useful not only for making future changes to eADVICE, but they will also be useful for assessing the connection between the health literacy and

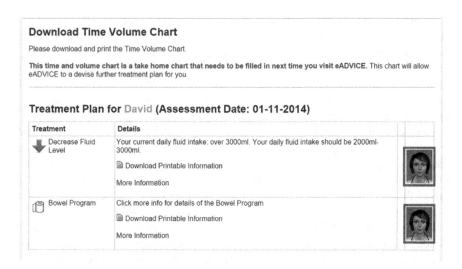

Download Time Volume Chart

Please download and print the Time Volume Chart.

This time and volume chart is a take home chart that needs to be filled in next time you visit eADVICE. This chart will allow eADVICE to a devise further treatment plan for you.

Treatment Plan for David (Assessment Date: 01-11-2014)

Treatment	Details	
▼ Decrease Fluid Level	Your current daily fluid intake: over 3000ml. Your daily fluid intake should be 2000ml-3000ml. 📄 Download Printable Information More Information	
🛢 Bowel Program	Click more info for details of the Bowel Program 📄 Download Printable Information More Information	

Fig. 5. Treatment Plan screen with link to Dr EVIE (right).

sense of therapeutic alliance in families to treatment adherence of patients in future work. The surveys were created using Qualtrics, which is an online research tool for generating, distributing and collecting results from surveys.

Two surveys were created for the study: prestudy and poststudy. This prestudy survey is taken by the patient's family before first use of eADVICE. The survey consists of four stages:

1. Patient Contact Details | provides details of the patient including contacts for referring clinicians
2. Enrollment Questionnaire | questions concerning patient's current condition and previous treatments
3. Quality of Life Questionnaire | questions asking the patient how their condition affects their life (e.g. Do you get embarrassed because of your bladder problem? with options Never, Sometimes, Often, All the time, I don't wish to answer).
4. Parent Perspectives and Literacy on eADVICE and Internet Questionnaire | questions that ask for parents' previous experiences using the internet for making decisions about health as well as questions assessing their health literacy.

The poststudy survey completed at the end of the trial consisted consists of an additional stage to assess the patient's experience with using eADVICE and the empathic agent. The format of this survey is described as follows:

1. Completion of Study Questionnaire | provides details of the patient including health outcome and treatments patient has been prescribed during the study.
2. Quality of Life Questionnaire | same as for pre-study survey.
3. Parent and Child perspectives of eADVICE including questions to determine if a working alliance has been established.
4. Usability Questionnaire that ask for parents' experience using eADVICE over the study period.

To measure the eHealth literacy of our participants we used the eHealth Literacy Scale (eHEALS) [23] as it is a short 10-item measure that has proven high reliability when tested on a youth population (aged 13–21). A 5 point Likert scale (strongly disagree to strongly agree) was used. Additionally, we included the Newest Vital Sign health literacy test that measures ability to interpret and apply health information by answering questions related to an ice cream label. The number of correct answers was summed to obtain a score.

To measure the extent of the therapeutic alliance with eADVICE we used the short version of the Working Alliance Inventory (WAI) [24]. This self-reported 12-item Likert scale questionnaire produces a composite score based on: WAI goal, WAI task and WAI Bond (which measures the degree to which the helper and client agree on the goals of the therapy and tasks to be performed, and the extent to which a bond between the helper and client is trusting and empathetic) using a 5-point Likert scale.

To be able to see if there is a relationship between eHealth literacy, therapeutic alliance and/or adherence we also measured reported adherence with the previous recommended treatment. Furthermore, the data from the Qualtrics surveys will be used to draw a connection between the use of the empathic agent and quality of life and health literacy of the patient.

5   Results and Discussion

5.1   Pre Study Results

Fourteen patients started the trial and did the prestudy survey consisting of 10 males and 4 females, aged 5–13 (M = 8.25). Only one family had used an online health program before. Table 1 shows the means for the 10 eHEALS eHealth literacy survey questions. The median and mode of all questions was 4.

Using the Newest Vital Sign health literacy test to produce a score of ability to apply health literacy skills, 9 participants scored 6/6, two scored 5/6 and one scored 1.5/6, indicating that all except one participant had good health literacy skills. Quality of Life scores ranged from 1.21–3/5 with mean of 2.093/5 with possible range 0–5.

5.2   Post Study Results

Thirteen (9 males and 4 females) completed the post study after the four month period, but one male reported being dry day and night and did not answer past the initial questions.

The poststudy survey inquired who had completed the survey. Seven surveys were completed by parent/s, and five were completed by the child and parent together. Quality of Life scores ranged from 1.21–2.6 with mean of 1.82/5. Four reported being dry day at night for 14 consecutive days, others had experienced day (4) and/or nighttime (10) wetting in that period.

We captured the patient history at the start of the trial (before) and at the end (after). Before the use of eADVICE, during the day three were wetting 4–7 times per week and another three 1–3 times per month, with six reporting no daytime wetting. For nighttime

Table 1. Heals eHealth literacy responses (1 = strongly disagree, 2 = disagree, 3 = neutral, 4 = agree, 5 = strongly agree)

Health literacy question	Mean
How useful do you feel the internet is in helping you in making decisions about your health?	4.17
How important is it for you to be able to access health resources on the internet?	4.33
I know what health resources are available on the internet:	3.75
I know where to find helpful resources on the internet	3.75
I know how to find helpful health resources on the internet:	3.92
I know how to use the internet to answer my questions about health:	3.92
I know how to use the health information I find on the internet to help me:	3.83
I have the skills I need to evaluate the health resources I find on the Internet:	4.00
I can tell high quality from low quality health resources on the Internet	3.83
I feel confident in using information from the Internet to make health decisions:	3.75

wetting, none of the patients had been dry for 10 consecutive days; seven had been wet more than once every night; four were wet 4–7 times per week and one was wetting 1–3 times at night per week. Exactly half of patients had tried treatments before. After the use of eADVICE, ten reported no daytime wetting, one reported less than once per month, another reported 1–3 times per week and two reported 1–3 times per week.

When asked "On average, how many times in the past 4 months has your child wet the bed/nappy?": four had been dry for 10 consecutive days; four more than once every night; three 4–7 times per week and three 1–3 times per week.

All respondents had tried one to five (M = 3.07) treatments since starting the study. Ten (10/12) were happy with the treatment and said their child was happy with the

Table 2. Results for Working Alliance Inventory questions (same scale as Table 1).

Working Alliance Inventory questions	Mean	SD
As a result of these sessions with eADVICE, I am clearer as to how I might be able to change	3.08	1.24
What I am doing with eADVICE gives me new ways of looking at my problem	3.16	1.11
I believe Dr Evie likes me:	3.75	1.06
Dr Evie and I collaborate on setting goals for my therapy	3.42	2.08
Dr Evie and I respect each other	4.25	0.75
eADVICE and I are working towards mutually agreed upon goals	4.08	1.16
eADVICE and I agree on what is important for me to work on	4.33	0.89
I feel that Dr Evie cares about me even when I do things that she does not approve of	3.83	1.03
I feel that the things I do in eADVICE will help me to accomplish the changes that I want	4	1.13
Dr Evie and I have established a good understanding of the kind of changes that would be good for me	4.16	0.94
I believe the way Dr Evie and I are working with my problem is correct	4	0.95

treatment. With the exception of one comment about a time and volume chart limitation and another stating *"too much details and work involve"* (by the person with poor health literacy score), further comments identified personal issues. For example,

Living in a rural area with a LOT of locums it was important for the program to give me info on medication rather than relying on a doctor. When it didn't do that I did not feel comfortable taking the medication [sic] option.

The responses to the Working Alliance Inventory and Usability questionnaires appear in Tables 2 and 3, respectively. Figure 6 shows the eight responses to the question "What were the most useful aspects of eADVICE?" Note the quote in bold.

When asked "What needed to be improved on eADVICE?" only one response concerned DrEvie *"It should take you automatically to the avatar rather then providing the link, some people might miss this"*. We made interaction with Dr Evie optional because we were interested to see how many people chose to use technology like Dr Evie and whether they would continue to use her. All patients had taken the link to Dr Evie at least once, the maximum number of times was 16 (M = 5.08, SD = 4.3,

Table 3. Usability/Usefulness questions. Legend: SD = Strongly Disagree, D = Disagree; A = Agree; SA = Strongly Agree; M = mean; SD = Standard Deviation.

Question	SD	D	A	SA	M	SD
I think that I would continue to use the eADVICE program to manage my child's wetting	0	3	6	3	3.75	1.14
I thought that eADVICE was easy to use	0	1	6	5	4.25	0.87
I felt confident using the eADVICE program	0	0	7	5	3.42	0.51
eADVICE is designed for all levels of users	0	0	11	1	4.08	0.29

1. *the connection that my daughter felt like the avatar was speaking to her; the timeliness of being able to find a treatment plan without waiting to see a Dr*
2. *Dr Evie for my daughter who is 6; A variety of treatment options some weeks; the language used.*
3. *Accessibility to information to help my child at home while waiting for an appointment at the Children's Hospital at XXX; Tailored information for my child and his needs; Ability to discuss with our GP treatment options recommended by the eAdvice and implement treatments under a medical practitioners guidance.*
4. *simple format*
5. *Bowel program*
6. *Ease of advice; Got a good feeling for the problem; Helped me know what to try*
7. *Dr Evie - good for the child to hear the advice from a non-family member. Computerised person appeals to the child. Info was presented very clearly, didn't leave us unclear about how to proceed.* **Dr Evie was encouraging, almost empathetic! It was good that someone pre-empted the likelihood that the program may not be followed completely/that well and that subsequent visits reinforced previous advice.**
8. *Convenience*

Fig. 6. The most useful aspects of eADVICE

median = 4). Use of eADVICE in this period ranged from 0–7 (M = 2.31, SD = 1.92, median = 2). Use of Dr Evie was thus greater than use of eADVICE and patients were visiting her between the fortnightly consultations with the eADVICE website.

6 Discussion

Since only one person had a very low score health literacy score, we are unable to answer whether the level of health literacy related to eADVICE and/or Dr Evie usage or satisfaction scores. Table 1 shows noteworthy health outcomes after using eAD-VICE over the 4 month period. It is important to remember that families in our study have struggled with the condition already for months and years before seeking spe-cialise help. The waiting list to see the specialist is up to 2 years long. From Fig. 5, we see that Dr Evie was specifically identified by 3/8 respondents as being most useful, others mentioned accessibility/convenience, tailoring, language used and information provided. In Table 2, we saw that all patients/families at least sometimes felt a working alliance with Dr Evie. When we compare the results of this study with previous studies on eADVICE without Dr Evie, we find greater adherence to the treatment advice, changing from around 50 % to 59 % overall, with adherence to caffeine reduction, increased fluid and bowel program treatments above 75 %. Note that if the patient becomes dry, they tend not to revisit the program to report this. Also, if patients visit 0–1 times, we cannot determine their treatment adherence – as eADVICE only asks for adherence at subsequent visits. If patients visit twice and were dry by the second visit – we cannot determine treatment adherence, because if they report they are dry, the program does not ask about adherence.

Similar to the work on a context-aware Hospital Buddy [25], we want to extend the current system to not only take into account the patient history and time and volume fluid data, but also to capture other types of data. One of the treatment options is use of an enuresis alarm that wakes the child if they wet the bed during the night. In another project, we have developed a mobile phone app that is used in conjunction with a mat with sensors that triggers the phone alarm. This data would allow us to record how many nights the alarm had been used and triggered, reducing the need to ask the patient. This would allow us to add some tailored and empathic dialogue, perhaps via Dr Evie or a text message sent to the parent/child from her.

We also have some concerns regarding whether children may be able to recognise that Dr Evie is not real and this may create other issues. In our trial, we excluded any patients with mental health issues. We also restricted the conversation to ensure that we did not elicit any content that would require mandatory reporting or which excluded the parent from the decision-making.

For our trials we did not want to have complex interaction or technology requiring the use of microphones or natural language. We chose text-based options as the means for patient/family interaction as being provided with selections can be less stressful and have an educative role [26], enabled acquisition of an accurate patient history that did not require training of speech recognition software or subject to speech recognition errors, and avoided issues around turn-taking.

7 Conclusion and Future Work

In 2015/6, we have been conducting a trial with 100 patients. So far 98 patients have commenced the six-month program, and 39 have completed. We will wait until all participants have completed the study before conducting data analysis.

We are in the process of changing the eADVICE interface so that Dr Evie will be used throughout the interaction, including for taking the patient history, and patients will not interact directly with the website at all. Further work on Dr Evie's facial expressions and body gestures might further improve the working alliance and additional animations could be developed to demonstrate some actions or procedures required by patients.

References

1. Buchanan, B.G., Shortliffe, E.H.: Rule-Based Expert Systems. Addison-Wesley, Reading (1984)
2. Langlotz, C.P., Shortliffe, E.H.: Adapting a consultation system to critique user plans. Int. J. Man Mach. Stud. **19**, 479–496 (1983)
3. Fenerty, S.D., West, C., Davis, S.A., Kaplan, S.G., Feldman, S.R.: The effect of reminder systems on patients' adherence to treatment. Patient Prefer Adherence **6**, 127–135 (2012)
4. Barbosa, C.D., Balp, M.-M., Kulich, K., Germain, N., Rofail, D.: A literature review to explore the link between treatment satisfaction and adherence, compliance, and persistence. Patient Prefer Adherence **6**, 39–48 (2012)
5. Vasbinder, E.C., Janssens, H.M., Rutten-van Mölken, M.P., van Dijik, L., de Winter, B.C., de Groot, R.C., Vulto, A.G., van den Bemt, P.M.: e-Monitoring of Asthma Therapy to Improve Compliance in children using a real-time medication monitoring system (RTMM): the e-MATIC study protocol. BMC Med. Inform. Decis. Mak. **13**, 38 (2013)
6. Martin, D.J., Garske, J.P., Davis, M.K.: Relation of the therapeutic alliance with outcome and other variables: a meta-analytic review. J. Consult. Clin. Psychol. **68**, 438 (2000)
7. DeVault, D., Artstein, R., Benn, G., Dey, T., Fast, E., Gainer, A., Georgila, K., Gratch, J., Hartholt, A., Lhommet, M., Lucas, G., Marsella, S., Morbini, F., Nazarian, A., Scherer, S., Stratou, G., Suri, A., Traum, D., Wood, R., Xu, Y., Rizzo, A., Morency, L.-P.: SimSensei kiosk: a virtual human interviewer for healthcare decision support. In: Proceedings of the 2014 International Conference on Autonomous Agents and Multi-agent Systems, pp. 1061–1068. International Foundation for Autonomous Agents and Multiagent Systems, Paris, France (2014)
8. Gratch, J., Lucas, G.M., King, A.A., Morency, L.-P.: It's only a computer: the impact of human-agent interaction in clinical interviews. In: Proceedings of the 2014 International Conference on Autonomous Agents and Multi-agent Systems, pp. 85–92. International Foundation for Autonomous Agents and Multiagent Systems (2015)
9. Lisetti, C.L.: 10 advantages of using avatars in patient-centered computer-based interventions for behavior change. SIGHIT Rec. **2**, 28 (2012)
10. Fitzgerald, M.P., Thom, D.H., Wassel-Fyr, C., Subak, L., Brubaker, L., Van Den Eeden, S. K., Brown, J.S.: Childhood urinary symptoms predict adult overactive bladder symptoms. J. Urol. **175**, 989–993 (2006)

11. Beun, R.J.: Persuasive strategies in mobile insomnia therapy: alignment, adaptation, and motivational support. Pers. Ubiquit. Comput. **17**, 1187–1195 (2013)
12. Baker, S., Richards, D., Caldwell, P.: Relational agents to promote ehealth advice adherence. In: Pham, D.-N., Park, S.-B. (eds.) PRICAI 2014. LNCS, vol. 8862, pp. 1010–1015. Springer, Heidelberg (2014)
13. Bickmore, T., Gruber, A., Picard, R.: Establishing the computer–patient working alliance in automated health behavior change interventions. Patient Educ. Couns. **59**, 21–30 (2005)
14. Greenley, R.N., Kunz, J.H., Walter, J., Hommel, K.A.: Practical strategies for enhancing adherence to treatment regimen in inflammatory bowel disease. Inflamm. Bowel Dis. **19**, 1534 (2013)
15. Bickmore, T., Ring, L.: Making it personal: end-user authoring of health narratives delivered by virtual agents. In: Safonova, A. (ed.) IVA 2010. LNCS, vol. 6356, pp. 399–405. Springer, Heidelberg (2010)
16. Schmid Mast, M., Hall, J.A., Roter, D.L.: Disentangling physician sex and physician communication style: their effects on patient satisfaction in a virtual medical visit. Patient Educ. Couns. **68**, 16–22 (2007)
17. Monkaresi, H., Calvo, R., Pardo, A., Chow, K., Mullan, B., Lam, M., Twigg, S., Cook, D.: Intelligent diabetes lifestyle coach. In: OzCHI Workshops Programme (2013)
18. Côté, J., Ramirez-Garcia, P., Rouleau, G., Saulnier, D., Gueheneuc, Y.-G., Hernandez, A., Godin, G.: A nursing virtual intervention: real-time support for managing antiretroviral therapy. Comput. Inf. Nurs. **29**, 43–51 (2011)
19. Pignone, M., DeWalt, D.A., Sheridan, S., Berkman, N., Lohr, K.N.: Interventions to improve health outcomes for patients with low literacy. J. Gen. Intern. Med. **20**, 185–192 (2005)
20. Zaragozá, I., Guixeres, J., Alcañiz, M., Cebolla, A., Saiz, J., Alvarez, J.: Ubiquitous monitoring and assessment of childhood obesity. Pers. Ubiquit. Comput. **17**, 1147–1157 (2013)
21. Klein, M., Mogles, N., van Wissen, A.: An intelligent coaching system for therapy adherence. IEEE Pervasive Comput. **12**, 22–30 (2013)
22. Dreyer, J.: Humanizing healthcare with technology. Streamlining the clinical documentation process augments doctor/patient interaction. Health Manage. Technol. **34**, 14–15 (2013)
23. Norman, C.D., Skinner, H.A.: eHEALS: the eHealth literacy scale. J. Med. Internet Res. **8**, e27 (2006)
24. Horvath, A.O., Greenberg, L.S.: Development and validation of the Working Alliance Inventory. J. Couns. Psychol. **36**, 223 (1989)
25. Bickmore, T., Asadi, R., Ehyaei, A., Fell, H., Henault, L., Intille, S., Quintiliani, L., Shamekhi, A., Trinh, H., Waite, K., Shanahan, C., Paasche-Orlow, M.K.: Context-awareness in a persistent hospital companion agent. In: Brinkman, W.-P., Broekens, J., Heylen, D. (eds.) IVA 2015. LNCS, vol. 9238, pp. 332–342. Springer, Heidelberg (2015)
26. Carnell, S., Halan, S., Crary, M., Madhavan, A., Lok, B.: Adapting virtual patient interviews for interviewing skills training of novice healthcare students. In: Brinkman, W.-P., Broekens, J., Heylen, D. (eds.) IVA 2015. LNCS, vol. 9238, pp. 50–59. Springer, Heidelberg (2015)

Short Papers

Improving Motivation in Survey Participation by Question Reordering

Rohit Kumar Singh[1], Vorapong Suppakitpaisarn[1,2(✉)], and Ake Osothongs[3]

[1] Department of Computer Science, Graduate School of Information Science
and Technology, The University of Tokyo, Tokyo, Japan
{rohitsingh544,vorapong}@is.s.u-tokyo.ac.jp
[2] JST ERATO Kawarabayashi Large Graph Project, Tokyo, Japan
[3] Department of Informatics, School of Multidisciplinary Sciences,
The Graduate University for Advanced Studies (SOKENDAI), Hayama, Japan
ake.osothongs@gmail.com

Abstract. We organized an experiment to show that survey participants take part more when the questionnaires started with less aggressive questions. In our earlier work, we used Bayesian probability and graph algorithms to find relative values of each personal attribute. Using that valuation, we created two sets of the questionnaire each differs in question order and ask 33 personal attributes from participants. The first set of the questionnaire ordered questions from personal attributes with high valuations such as passport number, driving license number, last name, and monthly income to personal attributes with low valuations such as nationality, gender and office country. On the other hand, the second set of questionnaire ordered from those with low valuations to those with higher valuations. As a result, the number of participants who received the second set of the questionnaire and agrees to submit some information is 71.42 % more than those who received the first set of the questionnaire. Moreover, the second set of participants spends much less time in filling the questionnaire, but provides 1.78 % more information on average. (Parts of contents in this paper is presented at the 3rd domestic meeting of JSAI Special Interest Group on Business Informatics (SIG-BI). Although the preprint version of this paper can be obtained at the conference website [15], the version is not refereed and not considered as a publication.)

Keywords: Innovative user interfaces · Privacy · Attribute valuation · Questionnaire design

1 Introduction

Data collection is one of the most important steps in the knowledge acquisition process. A company that can collect data with higher quality and quantity have more chance to have a better knowledge about its customers than a company that cannot. When some of the data is personal information, the collection is

© Springer International Publishing Switzerland 2016
H. Ohwada and K. Yoshida (Eds.): PKAW 2016, LNAI 9806, pp. 231–240, 2016.
DOI: 10.1007/978-3-319-42706-5_17

usually more challenging, since most of the collection participants do not want to disclose it [4]. Businesses need personal information not only to validate genuine users, but also for their future planning, product or service feedback, targeted advertisement and marketing as well as personal information trading with third parties. So, directly or indirectly all businesses need users' personal information to be competitive and cater their users more effectively and efficiently.

Because of the above reasons, personal information in this modern data driven age is critical asset for all businesses. Often businesses collect users' personal information through online registration forms [1]. For example, e-commerce sites, mailing services, social networks, micro-blogging sites, location, news and weather services, handheld devices etc. All of these required s to register before accessing their services.

Online questionnaire surveys are a quick and cost effective way of collecting data from a large number of people. But often, people feel insecure in providing their personal attributes and leave such questions unanswered. It significantly affects the overall response to the survey. For example, the unanswered questions are personal phone number, passport number, driving license number, email address, mobile phone number, date of birth, etc. Without these personal information attributes, companies are deprived of many business opportunities whereas, users are forfeit from better services and offers.

To lure the users, companies often provide monetary gifts in exchange of their personal information [8,13]. However, we strongly believe that, beside the gifts, there are techniques that can motivate participants' to reveal more information.

1.1 Our Contribution

In [9], it is shown that question order can affect a questionnaire's result. The result motivates us to wonder if the participants' motivation can be improved when the order is well chosen. Since it is known from psychological studies [3,5,6] that an individual is more likely to accept a large request, if he/she is asked to do a smaller request before. The results make us believe that participants get involved more when the survey questionnaire starts with less aggressive questions first.

We organize an experiment to show our hypothesis in this paper. From our previous work, which was based on Bayesian probability and graph algorithms, we defined the relative values for each personal attribute [11]. Considering questions that ask personal attributes with higher values as aggressive questions, we constructed two sets of the survey questionnaire. Each differs in question order and asked 33 personal attribute from survey participants. In the first set, we start the questionnaire with the most aggressive question such as passport number, driving license number, last name, and finish the questionnaire with personal attributes which are less aggressive and having a low valuation such as nationality, gender and office country (place of work). On the other hand, the second set of the survey questionnaire is ordered from questions asking attributes with low valuations to those asking attributes with high valuations.

By counting the number of users who agreed to participate in our survey in each experiment set, we find that 54.54 % of participants joined the second set of experiment, while only 31.87 % participated in the first. We can imply from the results that, by beginning the questionnaire with less aggressive questions, we can improve the participant ratio by 71.43 %.

Considering only the users who agreed to participate, we find that, in the first set of experiment, participants answer 66.42 % of questions asked. In the second set of experiment, 67.42 % of questions are answered. When the users who did not agree to participate are considered as users who answer zero question, 36.76 % of questions in the second set is answered, while only 21.06 % is answered in the first set.

Along with user data, we also record the time of entry to study participants response behavior for each set. By the time record, we find that the average time participants spent in the first set of experiment is 745.71 s, while the average time in the second set is 289.45 s. Together with the result in the previous paragraph, we can conclude that the second set of participants spends about thrice the least time in filling the questionnaire, but provides 1.78 % more information on average. This could mean that participants of the second set of experiment are motivated to complete the survey in a shorter time. Participants of the first set might get feared when they got aggressive questions at the beginning of questionnaires, and think more before giving any information.

1.2 Related Works

The Survey questionnaire order was studied since 1939 [14]. In [9], McFarland performs a study to understand the question order effect on questionnaire survey response. In the study, the questions are divided into two categories, general questions and specific questions. While general questions, ask survey participants, their general points of view on economics or politics, specific questions focus more on some specific topics. Examples of the questions include "In general, how interested would you say you are in politics?" or "How would you describe the current energy problem in the United States?". Examples of the questions include "Who is the most appropriate candidate for U.S. presidential election?" or "What is the cause of gasoline shortage?"

It is shown in the experiment that, when a general question are asked after few specific questions, the answer of the general question tends to be different from when it is asked before. For example, when few political questions are asked before the general question "In general, how interested would you say you are in politics?", the number of participants who answer "very interested" is much higher than when the general question is asked as the first question.

Instead of using easy specific questions, Bishop et al. [2] replace them with political questions that are too difficult for most participants. After the participants have been asked by the hard questions, the number of persons who specify themselves as "a person who interests in politics" gets smaller.

In [16,17], Tourangeau, Radsinski, Bradburn, and D'Andrade found that, when participants are asked if they agree with the abortion after questions about human right, they are likely to agree with it than when they are asked about the abortion after questions about morality.

Beside the works discussed in previous paragraphs, it is also found in other several works [12,18] that the question order can change survey results. That motivates us to believe that, besides changing the answers, having different question order can change the amount of information we can obtain from participants.

2   Research Methods

In this section, we describe how we are creating our survey questionnaire list based on the results in the previous section. In Subsect. 2.1 we describe methods used for data collection, and in Subsect. 2.2, we provide our questionnaire list.

2.1   Data Collection

In our previous works, we develop a technique for valuating each personal attribute. In [10], we conduct a survey on 532 Thai participants. The survey asks participants if they will "disclose" each of their 33 personal attribute when we provide a certain amount of incentive. The list of the personal attributes we consider in this research is provided in Table 1.

For data collection, we develop an online questionnaire using node.js[1], express.js[2] and HTML[3]. There are two instances of the web server running in parallel and hosting two different questionnaires. First set of questionnaire starts with high valuation personal attributes, i.e. "Passport Number" and ends with least aggressive questions about "Office Country". On the other hand, the second set of questionnaire starts with less valued personal attribute of "Office Country" and ends with high valued "Passport Number". We selected 44 people, most of them are students in The University of Tokyo for our online survey and invited through personal email id's. Half of the recipients receive the first set of questionnaire and another half receive a second set of questionnaire. The choice of survey to the recipient is randomly selected by us and respective survey link is provided in their invitation email.

We intentionally invite only persons who know us personally. That is an obvious confounding variable and impacts on the generalisability of the results, but can help participants avoid doubting about the organizers of the survey, or considering our mail as "Spam mail". We also assure in the invitation about users' privacy and data security. During the invitation process, we try as much as possible to give all participants the same information. The email content looks as given:

---

[1] https://nodejs.org/en/.

[2] http://expressjs.com/.

[3] http://www.w3.org/TR/html/.

Dear XXX,

We are a group of researchers at XXX Lab in Department of Computer Science (University of Tokyo). We are collecting users' personal information for our research.

We are happy to receive your complete input for our form, but feel free to leave the uncomfortable fields.

Survey Link: http://xxxxxx-xxx.net:1234/

Note: Collected data is only used for research and will not be shared with any third party.

2.2 Questionnaire Format

In [11], we also give attributes' values when the set of participants considered is limited to male and female. Since most of our survey participants are male, we selected to use the values obtained from male participants. Except two questions, we keep 31 questions similar to the results. We replace question "ID Number" and "Picture" with "Passport Number" and "Driving License Number" respectively. Previous study was organized in Thailand, where "ID Number" is provided to every citizen, but same is not available in Japan. Beside, we used "Driving License Number" instead of "Picture", because we assumed driving license is more accurate information about a person than its picture in an online survey about personal attributes.

We also use the same values for personal attribute disclosure(V_x) from the previous study [11]. For the simplicity of our research, we carry same V_x for the newly added attributes "Passport Number" and "Driving License Number" as calculated for existing attributes "ID Number" and "Picture". Our first set of questionnaire composes of sequence given in Table 1, and the second set is exactly the reverse of it so that it has a less aggressive question first. We invite people in our online survey on September 16, 2015 and collect data until September 23, 2015. A sample image of our online survey form is provided in Fig. 1.

3 Results Analysis and Discussion

To test our hypothesis that asking less aggressive question first increase the submission percentage in an online personal attributes questionnaire, we perform an online survey with two sets of the same questionnaire on 44 participants. Order of questions in set-1 is same as provided in Table 1, and set-2 is exactly reverse of set-1. 22 participants are invited to answer the order of questions in set-1, and the other 22 participants are invited to answer the order of questions in set-2.

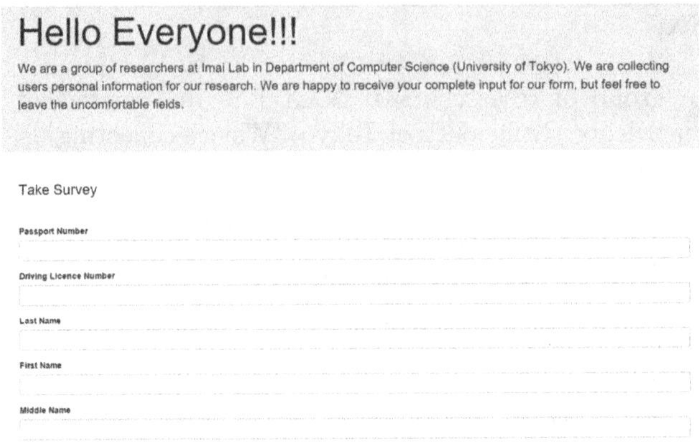

Fig. 1. Online Survey Form (Set 1)

As a result, 7 participants to set-1 and 12 participants from set-2 submits our survey. The data we receive from 44 invited participants, we calculate several statistical results as shown in Table 2. We use χ^2 test [20], Student's t-test [7] and Welch's t-test [19] to test the significance of these data.

We analyze the survey participation ratio for each questionnaire set. Since there is 7 submissions for set-1 and 12 submissions for set-2, we can conclude that there is an increase of 71.42 % more submissions for set-2. It also suggests that, online questionnaire survey on personal attributes which start with the least aggressive question first, encourages participants more for submission than questionnaire starts with the more aggressive question first.

To test the significance of the conclusion, we divide the participants in each set into two groups, participants who submit and participants who does not submit. By that, we can calculate p-value from the χ^2-test based on the technique discussed in [9]. Although, our p-value is as high as 0.127.

The average amount of time spent by participants in submitting set-1 and set-2 is 745 s and 289 s respectively, we can conclude that the time for set-2 is 61.18 % less than set-1. This implies that asking, most aggressive questions first, makes users more insecure for the whole survey and that lead to increase in time spend for the survey.

To calculate the average time in the above paragraph, we consider only participants who submit their information, that's why the numbers of participants in set-1 and set-2 are different. Because of that, we cannot test the significance of the test commonly used for quantitative data, Student's t-test. Instead of that, we use Welch's t-test [19] that is designed for this situation. The p-value we obtain from the test is 0.053.

The average number of personal attributes responded by participants for set-1 of questionnaire is 21.86 and set-2's average response is 22.25, we can conclude that there is a slight increase of 1.78 % for set-2 as compared to set-1. Similar to average time, we can use the Welch's t-test for testing the significance of this data. Although, the p-value obtained from the test is as high as 0.453.

Table 1. Survey Questionnaire Set

Rank	Attribute	VD
1	Passport Number (ID Number)	1.0000
2	Driving License Number (Picture)	0.9908
3	Last Name	0.9896
19	First Name	0.9430
21	Middle Name	0.9392
4	Home Address	0.9826
5	Home City	0.9826
6	Monthly Income	0.9789
7	Home Phone	0.9753
8	Office Email	0.9750
9	Highest Education	0.9692
10	Age	0.9583
11	Office Phone	0.9575
12	Marital Status	0.9540
13	Nickname	0.9504
14	Mobile Phone	0.9498
15	Personal Fax	0.9495
16	Number of Children	0.9490
17	Office Address	0.9444
18	Home Zip Code	0.9442
20	Home Province	0.9413
22	Birth Date	0.9383
23	Office City	0.9380
24	Office Zip Code	0.9377
25	Blood Type	0.9336
26	Personal Website	0.9302
27	Office Province	0.9297
28	Personal Email	0.9286
29	Home Country	0.9181
30	Nationality	0.9157
31	First Language	0.9111
32	Gender	0.8992
33	Office Country	0.8969

In above paragraph, we calculate the average attribute response value without considering persons who did not submit their information. Now, we consider them as persons who provide us 0 response. Using that, the average response of the set-1 is 6.95, and the average response of the set-2 is 12.13. We can conclude that set-2 improves set-1 by 74.53 %. Since the number of participants of set-1 and set-2 is not different for this data, we use Student's t-test for testing the significance. The p-value obtained from the test is 0.071.

Table 2. Comparison between Set-1 (Higher Valuation Attributes to Lower) and Set-2 (Lower Valuation Attributes to Higher)

Average Value	Set-1	Set-2	Significance test	Significance
Participation ratio	32 %	55 %	χ^2-test	0.127
Average time taken for form submission	745s	289s	Welch's t-test	0.053
Average number of attributes filled (considered only users who agree to participate)	21.86	22.25	Welch's t-test	0.453
Average number of attributes filled (considered users who do not participate as users who give us zero attribute)	6.95	12.13	Student's t-test	0.071
Average number of times distracted	2.86	2.16	Welch's t-test	0.326

The average number of questions for which participants spent more than 30 s to answer, We consider it as an distraction. Participants in set-1 distract from our survey for 2.86 times on average, while participants of set-2 distract for 2.16 times, on average. The number of times of distraction in set-2 is 24.48 % smaller than the number in set-1. We may be able to conclude from there that the participants in set-2 are more motivated than the participants in set-1. Again, we use Welch's t-test to test a significant. The p-value obtained from the test is as high as 0.326. We can conclude that there's insufficient evidence to reject the null hypothesis that there's no effect.

All above five results, show improvements for set-2 in comparison to set-1. In addition, at the significance level of 0.10, the result of "Average time taken" in Table 2 is significant. Moreover, at this significance level, the result obtained for "Average number of attribute filled" in Table 2, with the consideration of total invitees for each set, i.e., 22, is also significant. However, significance test does not support our improvements at a significance level of 0.05 or 0.01. We believe that, for a justifiable significance test, a large number of participants is crucial. However, obtaining such a critical number of participants is very difficult for these kind of experiments.

We have shown in previous paragraphs that set-2 have a larger number of responses because of a higher motivation of participants. However, in many applications, the information that we really want is the most variable information

Table 3. Top 5 Significant Attributes

	Passport Number	Last Name	Home Address	Home City	Monthly Income
Set-1	1	7	6	7	6
Set-2	3	8	7	11	9

for participants' point of view. In Table 3, we show the number of participants answering the top 5 most significant attributes from both sets of question-naires. These are, *Passport Number, Last Name, Home Address, Home City, and Monthly Income.*

Out of 7 respondents for set-1 and 12 respondents for set-2, very few have revealed their *Passport Number*. However, *Home City* and *Monthly Income* is answered by most of them. In addition, *Last Name* and *Home Address* is almost equally answered by both groups of respondents. Despite the difference in their ordering, in our opinion, the reason for not having any big difference in the response for each set of above high value attributes are; for set-1, participants are more careful in revealing their most important information; whereas, in set-2 participants are already exhausted in providing initial least aggressive questions and lost their motivation for disclosing more valuable personal attributes.

4   Conclusion

For this study, we adopt a psychological technique called "foot in the door" [3,5,6]. In this technique, by answering the least aggressive questions first, par-ticipants are obligated to answer later comparatively higher valuation aggres-sive questions. Our result analysis reveals that set-2 questionnaire submission is approximately seventy percentage higher than set-1. Moreover, the average time spend by participants on submission of set-2 is around sixty percentage less than set-1. These results confirm our previous study of personal attribute valuation. It also supports our claim that participants respond more in an online questionnaire when it starts with less aggressive questions first.

However, significance test supports our claim partially. We believe that, with a large set of participants our results significance could improve. As a result, for future works, we are planning to replicate our experiment on a large population to justify our results significance. We also plan to conduct this experiment on Thai participants to avoid cross-cultural factors between Thailand and Japan. In this paper, we use fixed values for attributes. However, we can dynamically adjust the personal attribute value based on user responses for a particular attribute. After we adjust those values, we can adjust the question order based on the values, and finally we might obtain the optimal order.

Acknowledgement. The authors would like to thank Mr. Sra Sontisirikit and Mr. Ratthachai Chawuthai for ideas that initiate this research. We also would like to thank Prof. Takao Terano, who give us several valuable comments after we made a presentation at the 3rd domestic meeting of SIG-BI

References

1. Awad, N.F., Krishnan, M.S.: The personalization privacy paradox: an empirical evaluation of information transparency and the willingness to be profiled online for personalization. MIS Q. **30**(1), 13–28 (2006)
2. Bishop, G.F., Oldendick, R.W., Tuchfarber, A.: What must my interest in politics be if i just told you "i don't know"? Public Opin. Q. **48**(2), 510–519 (1984)
3. Burger, J.M.: The foot-in-the-door compliance procedure: a multiple-process analysis and review. Pers. Soc. Psychol. Rev. **3**(4), 303–325 (1999)
4. Chellappa, R.K., Sin, R.G.: Personalization versus privacy: an empirical examination of the online consumers dilemma. Inf. Technol. Manag. **6**(2–3), 181–202 (2005)
5. Freedman, J.L., Fraser, S.C.: Compliance without pressure: the foot-in-the-door technique. J. Pers. Soc. Psychol. **4**(2), 195 (1966)
6. Gueguen, N., Meineri, S., Martin, A., Grandjean, I.: The combined effect of the foot-in-the-door technique and the "but you are free" technique: an evaluation on the selective sorting of household wastes. Ecopsychology **2**(4), 231–237 (2010)
7. Haynes, W.: Students t-test. In: Dubitzky, W., Wolkenhauer, O., Cho, K.-H., Yokota, H. (eds.) Encyclopedia of Systems Biology, pp. 2023–2025. Springer, New York (2013)
8. Kroft, S.: The data brokers: Selling your personal information. http://www.cbsnews.com/news/data-brokers-selling-personal-information-60-minutes/. Accessed 22 October 2015
9. McFarland, S.G.: Effects of question order on survey responses. Public Opin. Q. **45**(2), 208–215 (1981)
10. Osothongs, A., Suppakitpaisan, V., Sonehara, N.: Evaluating the importance of personal information attributes using graph mining technique. In: Proceedings of the 9th International Conference on Ubiquitous Information Management and Communication (IMCOM 2015), No. 104, ACM Digital Library (2015)
11. Osothongs, A., Suppakitpaisan, V., Sonehara, N.: A proposed method for personal attributes disclosure valuation: a study on personal attributes disclosure in Thailand. In: Proceedings of the 9th International Conference on Information Technology and Electrical Engineering (ICITEE 2015), pp. 409–413. IEEE Xplore (2015)
12. Pew Research Center: Questionnaire design: Question order. Technical report, Pew Research Center, U.S. Politics & Policy (2013)
13. Phelps, J., Nowak, G., Ferrell, E.: Privacy concerns and consumer willingness to provide personal information. J. Public Policy Mark. **19**(1), 27–41 (2000)
14. Sayre, J.: A comparison of three indices of attitude toward radio advertising. J. Appl. Psychol. **23**(1), 23 (1939)
15. Singh, R.K., Suppakitpaisarn, V.: Improving motivation in survey participation by question reordering. In: the 3rd meeting of JSAI Special Interest Group on Business Informatics (SIG-BI), No. 8 (2015)
16. Tourangeau, R., Rasinski, K.A.: Cognitive processes underlying context effects in attitude measurement. Psychol. Bull. **103**(3), 299 (1988)
17. Tourangeau, R., Rasinski, K.A., Bradburn, N., D'Andrade, R.: Carryover effects in attitude surveys. Public Opin. Q. **53**(4), 495–524 (1989)
18. Weisberg, H.F.: The Total Survey Error Approach: A Guide to the New Science of Survey Research. University of Chicago Press, Chicago (2009)
19. Welch, B.L.: The generalization of student's problem when several different population variances are involved. Biometrika **34**(1/2), 28–35 (1947)
20. Yates, F.: Contingency tables involving small numbers and the χ^2 test. Suppl. J. R. Stat. Soc. **1**(2), 217–235 (1934)

Workflow Interpretation via Social Networks

Eui Dong Kim and Peter Busch[✉]

Department of Computing, Macquarie University,
Sydney, NSW 2109, Australia
peter.busch@mq.edu.au

Abstract. We sought to determine how people worked in practice, how management saw they worked and examine 'gaps' between these two 'views'. In order to see potential differences, we examined workflow management through interviews with managers and a questionnaire with employees. The results were analysed through Petri Nets in a simplified form. The second unit of analysis was examining relationships between employees and therefore their knowledge flows using Social Network Analysis to illustrate work patterns staff had with one another. Through overlaying the two we gained some understanding of matches and mismatches. The study took place in three IT units of one organisation – an Australian university. The outcomes of our study comprise potential recommendations for improving work efficacy, such as re-organising work practices, or potentially changing who works with whom.

Keywords: Workflow · Formalisation · Petri nets · Business process modelling · Social network analysis · Business process management · Knowledge management

1 Introduction

Work or business processes are usually set by management, and so tends to be structured from management's viewpoint. However, no one can be sure the actual work performed is along management's requirements. We seek to determine how people work in practice, how management sees they work and then examine the 'gaps' between these two 'views' of work. In order to see these potential differences, we adopt process management which captures work flows as well as Social Network Analysis (SNA) which illustrates relationships staff have with one another. Through overlaying the two we gain some understanding of mismatches in work patterns. Some recommendations are provided with regard to gaps found for improving business.

2 The Human Aspect of BPM

There is much literature which deals with Business Process Management [1], but most of this concerns task coordination. There has also been some work on the human aspect of process management [2–5]. Generally however overlaying human relationships with 'official' working relationships is less well understood. Analysing existing social networks in the organisation supports business developers in designing, developing and

H. Ohwada and K. Yoshida (Eds.): PKAW 2016, LNAI 9806, pp. 241–250, 2016.
DOI: 10.1007/978-3-319-42706-5_18

implementing business processes more efficiently and effectively. As early as the 1990s, O'Reilly [6] had used five themes to explain such impacts from different aspects – being motivation, leadership, job design, turnover/absenteeism and "work-related attitudes, such as job-specification, as affective evaluations about aspects of one's work environment" (p. 435). Wasserman and Galaskiewicz [7] added another theme - power. Employees may take on similar attitudes with whom they co-operate or to individuals who occupy the same positions in the social network. The value of examining activities of actors/employees through Social Network Analysis is that we can examine the mapping of tasks to personnel at a relatively low level of granularity, asking questions of the relationship of staff with their work processes, or of their working relationships with other colleagues. It is an assumption of SNA that relationships between people act as thoroughfares of resources.

3   Social Network Analysis

We may define "Social Network Analysis as an established social science approach of studying human relations and social structures by disclosing the affinities, attraction and repulsions operating between persons, and between persons and objects" ([8] p. 64). SNA provides the researcher with information showing what promotes or interrupts knowledge flows, through showing "who knows whom, whom shares what information and knowledge with whom by what communication media" ([9] p. 2). There are several advantages to applying SNA to workflows, such as providing a more seamless customer service [10], better understanding the unseen roles of workers [11], or improving workflow control mechanisms [12]. Ideally just observing social networks is not in itself completely meaningful, for observations should be compared with pre-defined business processes known as workflows [13].

4   Methodology

Three case study groups within the same Australian university were conducted. Numerous internal and external documents were used as data sources. Interviews with IT managers of each unit was a means of data gathering. Interview questions comprised (1) what kind of jobs does this department handle? What are the roles and responsibilities? (2) How many people work in this department? Please tell me their names. (3) Do you have an organizational chart? - are there sub-divisions or teams? (4) what are the major business processes in this department? (5) can I have a job description for each position? (6) Do you use an information system to manage business processes? (6a) what is it? Does that system provide a view of business processes? Data gathered through interviews was analysed through UCINET SNA software, which produced sociograms and matrices and enabled us to visualise centrality, density, inclusiveness, and cliques via analysing degree, closeness and betweenness. WoPeD (Workflow Petri Net Designer) was used for designing Petri-nets. The idea of SNA graphs was to illustrate relationships between personnel (typically bi-directional). Using Petri nets allowed (often uni-directional) workflows to be modelled.

5   Case Sites

Unit A (Fig. 1a) is an university library IT department with 10 staff including a manager. This unit with seven employees including one IT operations manager, provides IT services to users of the library. Another section is design and development comprising two senior systems analysts. Unit B (Fig. 1b) is an IT service department with eleven staff including a manager, providing IT services to staff and students in the faculty and is largely divided in to two parts. One is a service desk team which normally has five employees, including one service desk manager. The other is system administration with six employees. Unit C (Fig. 1c) is an IT service department of another faculty with six staff including a manager, providing IT services to staff and students in the faculty.

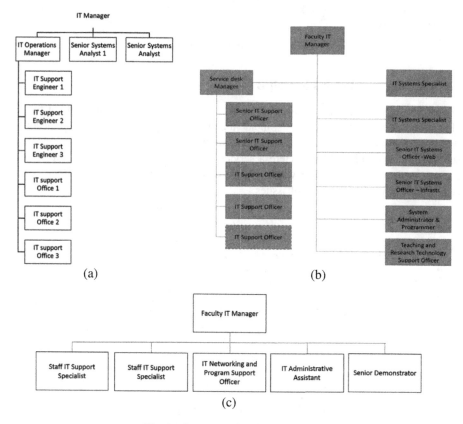

(a)   (b)

(c)

Fig. 1. Structure of Unit A, B and C

6　Workflow Results

Each unit manager was asked to choose a certain business process in which all of their employees were involved as there is more than one business process in a department, and not all employees are related to all processes - even within the same department. Business processes chosen were - an incident management process, a service desk process and a software deployment process.

6.1　Incident Management Process in Unit A

An incident can be defined as an unplanned interruption to an existing IT Service or reduction in the quality of an IT Service. The incident management process can be drawn hierarchically as it has certain sub-processes, although here we show just the top process for space reasons (Fig. 2).

Fig. 2. High level incident management process

6.2　Service Desk Process in Unit B

Unit B chose a service desk process. A service desk is a one stop contact point for its clientele, including service requests, incidents, problems, providing advice, an account or access to a server. End users or customers might not be able to distinguish incidents, service requests and problems. Here a service desk business process was drawn at a conceptual level (Fig. 3), as it would be too complicated to break down the process according to the classification of issues.

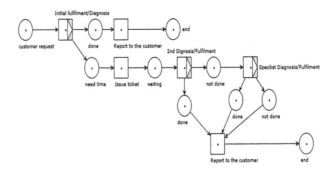

Fig. 3. Service desk business process

6.3   Software Deployment Process in Unit C

Again we show just the top level (Fig. 4) of a software deployment process which starts with a user's request for a new application, followed by checking whether it is available in the Standard Operating Environment (SOE). If it is available in the SOE, a licence is required for which there are two types - a site licence, or one for individuals.

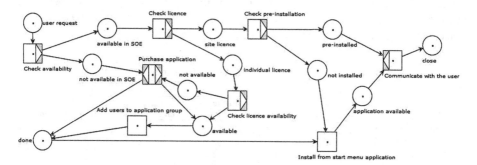

Fig. 4. A software deployment business process

If a site licence (whether pre-installed or not), it is checked. If it is pre-installed, the user can access software, so the work-case is closed. If it is not installed, installation is performed from the start menu, so the application becomes available for the user - the workcase closes. In the case of an individual licence, licence availability is also checked and if it is not available, it goes to the purchase application. If it is available, users are added to an application group and some work is done for installation from the start menu application. Given space limitations, let us now discuss the SNA results.

7   SNA Results

In the aforementioned questionnaire, levels of frequency of contact, importance of the staff member, and nature of contact were analysed. There were eight responses out of nine from unit A. In unit B, eight people responded out of eleven employees. Also four out of five people participated from unit C. The questionnaire was distributed to all employees, except managers as their viewpoints had been captured through interviews.

7.1   SNA in Unit A

Table 1 shows a sociomatrix which is converted in to the sociogram (Fig. 5) representing frequency of contact in unit A. The integers in this sociomatrix range from one refering to contact once every three months (quarterly), to nine signifying hourly contact - the highest frequency of contact. The eight participants are labelled from actor one to actor eight for reasons of anonymity. The width of ties between the nodes represent the strength of relationships among actors. We see actor two has a stronger relationship with actor seven than with any other actors. Also, actor 1 has a strong overall relationship with all other actors. The employees of the IT operations team in organization A belong to the first clique or the third clique. The second clique includes two employees from the design and development team, and one employee from the IT operations team. Here we surmise actor four can play a coordinating role between the IT operations team and the design and development team. In the IT operations team, actors five and six have a closer relationship with the manager of the IT operations team. Also, actor five has the widest relationship with all employees in the IT operations team, while actor 8 appears to be relatively isolated.

Table 1. Sociomatrix showing contact frequency in unit A.

		Frequency of contact							
		Actor 1	Actor 2	Actor 3	Actor 4	Actor 5	Actor 6	Actor 7	Actor 8
Name	Actor 1	0	8	8	8	8	8	8	8
	Actor 2	5	0	3	3	3	3	6	2
	Actor 3	8	6	0	9	8	9	6	8
	Actor 4	8	7	9	0	8	9	7	8
	Actor 5	8	6	8	8	0	8	6	8
	Actor 6	9	6	9	9	7	0	6	7
	Actor 7	7	8	6	6	5	6	0	5
	Actor 8	9	8	9	9	9	9	8	0

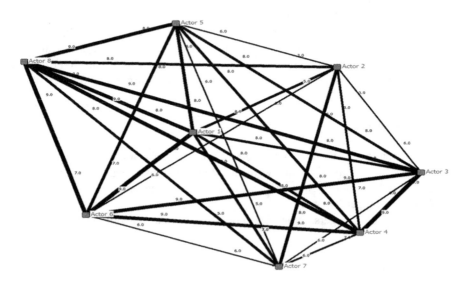

Fig. 5. Sociogram of contact frequency in unit A.

Similar sociomatrices and sociograms were conducted for importance of person in each case unit where the integers ranged from one to five, where five meant 'must see the person' - the highest importance of the person, to one meaning 'no need to see the person at all' - the lowest importance of the person. Again the same process was followed for 'importance of the occasion' which was measured as an integer from one to eight. In this case eight meant a routine or formal meeting - defined as the highest importance of the occasion, where one meant 'usually see the person outside of work' - defined as the most informal occasion, namely the lowest importance of the occasion.

In SNA we can also measure centrality and power through three concepts: degree, closeness and betweenness. In terms of degree, both actors four and five have a degree of five meaning they have five ties with other actors respectively. In terms of closeness actors four and five possess the shortest path length from other actors (Table 2). With regard to betweenness, both actors four and five are between six other pairs of actors respectively, therefore both actors four and five have what is referred to as centrality and power in unit A. We may also state the density in unit A is one hundred percent and the inclusiveness is also one hundred percent, meaning all actors are connected and included – no-one is isolated. These figures can be different if we adjust the value of the threshold or edges in the graph or sociogram for each organisational unit. Using the same techniques which we cannot illustrate here for reasons of brevity, the eight participants from unit B, showed actor H had a strong overall relationship with all other actors. Actor H in particular had a stronger relationship with actors A and B than with any other actors.

Table 2. Closeness between actors in unit A

Name	Path length between actors								
	Actor 1	Actor 2	Actor 3	Actor 4	Actor 5	Actor 6	Actor 7	Actor 8	Sum of path length
Actor 1	0	3	2	2	1	1	3	2	14
Actor 2	3	0	2	1	2	2	1	3	14
Actor 3	2	2	0	1	1	1	2	2	11
Actor 4	2	1	1	0	1	1	1	2	9
Actor 5	1	2	1	1	0	1	2	1	9
Actor 6	1	2	1	1	1	0	2	2	10
Actor 7	3	1	2	1	2	2	0	3	14
Actor 8	2	3	2	2	1	2	3	0	15

Actor H was a manager of the service desk team, so he had a role to allocate service desk jobs to the appropriate employee, which is why he had a closer relationship with all other employees in unit B. Also, even actors A and B belonged to his team, so the frequency of contact among them was necessarily higher than with other employees.

Again the same procedure was used in examining the SNA results for the 4 actors in unit C. Remember not every employee chose to participate in the SNA question-naire, and university human research ethics committees (HRECs) specifically prevent this sort of coercion in a research context. Again, given space limitations we cannot delve deeper into the techniques but will next present our general findings of relevance to knowledge acquisition and management.

8 Discussion

Through using petri nets to examine work processes and using SNA to explore working relationships between employees in the three different units, we were able to generalise our findings, although we acknowledge this is work in progress.

Unit A had only ten employees including a manager, so the overlap between management's viewpoint and actual working relationships was relatively large. The frequency of contact amongst employees was high overall as they often saw each other at formal and informal meetings. Also workers in the same group had closer relationships than with workers in other groups - meaning their social relationships were almost fully overlapping with management's viewpoint. However, some hidden relationships were found. There were three cliques not fully overlapping in the organization structure. These cliques may ideally be utilized as work groups or teams when there is a need for a specific project, a task force team, or re-organization. Also two workers who are not managers, had centrality and power so they could potentially play a coordinating role and be utilized as leaders or a sub-managers of small groups or teams. Another factor to be considered was one slightly isolated employee who potentially needed to be mentored carefully to see if his isolation caused bottlenecks or pain points in any given business process.

Unit B was also small having twelve employees including a manager. One finding was that the relationships among three actors in the same team (the service desk team), was stronger than the relationship among other actors. This finding was determined by overlapping the organization chart with the SNA results. Also 'was there a difference in SNA between unit A and unit B? In regard to the importance of the occasion, the SNA values in unit A were mostly high, such as seven or eight meaning strong contact, but the values in unit B were somewhat lower (collaborating but not formally) and different from one another. As such it would appear unit B is typified by more informal meeting types than unit A, even though the size of organization is nearly the same. Unit A is more tightly structured with a team, however several employees in organization B are not tied in to a team, which could be a possible cause of the differences between the two SNA analyses. There were also certain cliques (groups of actors >= 3), which were invisible from management's point of view. Also, the person who has the most centrality and power - actor H, was a service desk manager, but the employee who had the second highest centrality and power - actor C, belonged to cliques which did not include actor H; meaning actor C is a key employee who ideally needed to cooperate with actor H.

Unit C had five employees, the smallest among the three case sites. There are no teams or working groups within this unit. Overlaying BPM with SNA, there were few differences between management's viewpoint and actual social relationships. However, there were also two cliques which were not seen from management's perspective. These two cliques could be utilized as working groups or teams. Also one employee formed the centre of a social network - belonging to both cliques; this actor could be utilized as a team leader or sub-manager, playing a coordinating role between the manager and employees. Another parameter to be considered was that actor L had a relatively weak relationship with actor J; perhaps there were conflicts between them?

Or they may have felt less comfortable when requiring cooperation from another. If there is a need for people to work together, it is necessary to consider the social relationship between them in order to improve work efficacy.

9   Limitations

For ethics committee approval reasons we can not identify individual employees so the relationships depicted tend to be at 'arms length'. During the questionnaire, one question was asked continually by employees of each of the case units under study - why managers of each unit were not included in the questionnaire? In the SNA component of the questionnaire, managers were not in the list of employees. This was because managers and their employees needed to be separated to clearly distinguish management's viewpoint from the social relationship amongst employees themselves.

Also worth mentioning was that the scope of the research was not university-wide - it was limited to selected units within one university. Because chosen business processes were all internal, meaning there were no cross-departmental or cross-sector workflows, the impact of social relationships on business processes were also limited. Staff sat close to each other, met every day, and knew each other's work very well. If a workflow is a cross departmental one, social relationships with other department's employee's would vary and likely give more distinctive results. We acknowledge these are limitations in this pilot study.

Also many employees had the same flow of work but worked independently. There was no work which was performed by one employee exclusively. So, it was difficult to map each task exclusively to a particular employee. It is very important to select a suitable business process to be analysed which is clearly defined, the tasks of which are allocated to each employee independently; otherwise it is difficult to map each employee to the drawn business process. Since a business process is typically a broader view pre-defined by management, it may not be decomposed to the level of detail required. Therefore at such a high level there may appear to be no mismatches in work patterns. In future research, detailed and appropriately decomposed level of workflows need to be derived and analysed. Also cross-departmental business flows which show clear roles and duties of each employee, need to be chosen ideally in a larger organization that is willing of course to participate.

10   Conclusion

This study neither explained a causal relationship nor generalized a theory, but tried to explore a social phenomenon. Case studies with three units within the one overlying - a university, were designed and conducted through interviews and a questionnaire survey. The research examined the overlaps of workflows with Social network Analysis (SNA). Organizational structures for each unit were drawn from data gathered during interviews with managers of the case units and used as a representation of management's view. For workflow analysis, simplified Petri Nets were used to understand the workflow of each organization. For SNA, sociograms, sociomatrices and other quantitative measurements (such as centrality, betweenness, prestige and closeness) were used to analyse actual working relationships in the case units. Management's viewpoint through petri nets as well as organization structure charts were compared with SNA findings, which showed working relationships amongst employees. The result showed some gaps between the two, but the gaps were not large as the number of employees in the case organizations were small. Nonetheless, some helpful findings was provided to management from gaps discovered. More importantly, this study trailed certain research approaches and determined weak points, limitations, and issues to be considered, which could be meaningful for refining such techniques in the future. Hopefully the results, methodologies, and experiences of this study can be shared with other researchers to improve the quality of further research.

References

1. Houy, C., Fettke, P., Loos, P.: Empirical research in business process management-analysis of an emerging field of research. Bus. Process Manag. J. 16(4), 619–661 (2010)
2. Deokar, A., Kolfschoten, G., de Vreede, G.: Prescriptive workflow design for collaboration-intensive processes using the collaboration engineering approach. Global J. Flex. Syst. Manag. 9(4), 11–20 (2008)
3. Fisher, D.: The business process maturity model: a practical approach for identifying opportunities for optimization. Business Process Trends (2004). www.bptrends.com (accessed 12 Dec 2015)
4. Magdaleno, A., Cappelli, C., Baião, F., Santoro, F., Araujo, R.: Towards collaboration maturity in business processes: an exploratory study in oil production processes. Inf. Syst. Manag. 25(4), 302–318 (2008)
5. Rosemann, M., de Bruin, T., Power, B.: A model to measure business process management maturity and improve performance. In: Business Process Management. Butterworth-Heinemann (2006)
6. O'Reilly, C.: Organizational behaviour: where we've been, where we're going. Ann. Rev. Psychol. 42, 427–458 (1991)
7. Wasserman, S., Galaskiewicz, J.: Politics and organisations. In: Advances in Social Network Analysis. Sage Publications (1994)
8. Hassan, N.: Using social network analysis to measure IT-enabled business process performance. Inf. Syst. Manag. 26(1), 61–76 (2009)
9. Serrat, O.: Social network analysis. In: Knowledge Solutions. Asian Development Bank Mandaluyong, Philippines (2009)
10. Bonchi, F., Castillo, C., Gionis, A., Jaimes, A.: Social network analysis and mining for business applications. ACM Trans. Intell. Syst. Technol. 2(3), Article 22, 37 pages (2011)
11. Poltrock, S., Handel, M.: Modeling collaborative behavior: foundations for collaboration technologies. In: Hawaii International Conference on System Sciences (HICSS), pp. 1–10 (2009)
12. Harrison-Broninski, K.: Dealing with human-driven processes. In: Rosemann, M. (ed.) Handbook on Business Process Management 2. International Handbooks on Information Systems, pp. 443–461. Springer, Heidelberg (2010)
13. Papazoglou, M., Ribbers, P.: e-Business: Organizational and Technical Foundations. John Wiley & Sons Ltd., Chichester (2006)

Integrating Symbols and Signals Based on Stream Reasoning and ROS

Takeshi Morita[✉], Yu Sugawara, Ryota Nishimura, and Takahira Yamaguchi

Faculty of Science and Technology, Keio University, 3-14-1 Hiyoshi,
Kohoku-ku, Yokohama 223-8522, Japan
{t_morita,nishimura,yamaguti}@ae.keio.ac.jp
http://www.yamaguti.comp.ae.keio.ac.jp

Abstract. We have developed PRactical INTElligent aPplicationS (PRINTEPS) which is a total intelligent application development platform. This paper introduces an application of PRINTEPS for detecting events by using stream reasoning and Robot Operating System (ROS), and for integrating image sensing with knowledge processing. Based on this platform, we demonstrate that the behaviors of a robot in a robot cafe can be modified by changing the applicable rule sets.

Keywords: ROS · Stream reasoning · PRINTEPS · Ontology

1 Introduction

We are currently developing PRactical INTElligent aPplicationS (PRINTEPS), a platform for developing a total intelligent application that achieves task collaboration between a machine and human by simply reconfiguring software modules that govern knowledge based reasoning, spoken dialogue understanding, image sensing, manipulations, and machine learning [8].

Designing machine-human task collaboration often requires integration of the image sensing technologies that help recognize surrounding circumstances by using a rule set of a target operation. However, a major hurdle exists in connecting the two directly. This is because a huge grain-size difference exists between information acquired through image sensing and that expressed by a rule set.

Since business rule management systems (BRMS) [2] have received much attention lately, increasing number of business people are developing systems that utilize business rules. BRMS is a framework that effectively uses business rules and it can flexibly manage changes of those rules.

However, one problem is that a robot cannot directly handle such business rules. A robot perceives differently from a person. There are various types of image-sensing software, which are employed as tools for robots to perceive objects and people [4]. Thus, a platform is needed that can link the image-sensing software programs and their business rules, and that can easily respond to changes of such rules.

© Springer International Publishing Switzerland 2016
H. Ohwada and K. Yoshida (Eds.): PKAW 2016, LNAI 9806, pp. 251–260, 2016.
DOI: 10.1007/978-3-319-42706-5_19

In other words, in order to map a rule set to the real world, we must not only convert physical information obtained through image sensing into certain signals but also have a platform that links those signals with the rule set.

This study, having a primary objective of mapping a group of verbally defined service rules to the real world, focuses on the link between a rule set and signals. It does not focus on image sensing and its details.

As a means to achieve integration between rules and image sensing, we propose a technique using certain languages for stream reasoning [3], continuous simple protocol and Resource Description Framework (RDF)[1] query language (C-SPARQL) [1], and Robot Operating System (ROS) [5]. C-SPARQL is a language that extends SPARQL and it enables event processing using RDF stream that employs time-stamped RDF triples. This technique allows a human operator to consider events based on those detected by C-SPARQL rather than based on an image sensing result. This means the rule set owned by the human operator is usable by the robot, thus further enriching robot services.

We conducted a case study of a teahouse customer-reception service using Pepper[2], which is an emotion-recognizing humanoid robot jointly developed by Aldebaran Robotics and SoftBank Mobile.

2   Integration of Symbols and Signals in PRINTEPS

2.1   Use of ROS in PRINTEPS

ROS [5], which is a robot framework developed by Open Source Robotics Foundation, offers communication libraries and various tools. In ROS, a user can employ Service, which has the same synchronous communication mechanism as the Web service, and Topic, which offers a mechanism for asynchronous communications. Because ROS is implemented as a distributed system, its use thereof facilitates the handling of multiple robots or multiple sensors, and the use of Topic enables various other processing operations, including the realtime acquisition of sensor values. The input-output data type of Service and Topic, known as Message, can be defined as a composite data type such as a class or structure. Regarding Service, the input-output data specifications can be described in srv files, without relying on a programming language. Currently, on the Wiki page related to ROS, more than 2,000 ROS packages have been released to the public [3], and various libraries for images, sensors, and motions are being actively developed by researchers around the world based on the Topic or Service specifications of ROS.

For the aforementioned reasons, we decided to use ROS Topic and Service as the minimal-function modules of PRINTEPS. Thus, we not only are able to reuse the many existing libraries, but can also solve problems such as the varying input-output data structure for each unit of intelligence as well as each instance of realtime processing through the use of Topic and Service.

---

[1] http://www.w3.org/RDF/.

[2] https://www.aldebaran.com/en/cool-robots/pepper.

[3] http://www.ros.org/browse/list.php.

2.2 Use of Stream Reasoning in PRINTEPS

This study uses stream reasoning as a means to link knowledge inference with other intelligence. Stream reasoning [3] is an academic disciplinary area for providing techniques or tools necessary to integrate data stream, semantic Web, and reasoning systems and thus provide complex and intelligent decision-making based on stream data input by a sensor.

Stream reasoning converts sensor data into time-stamped RDF data (RDF stream) and enables event processing through a retrievable language with extended SPARQL for searching an RDF database. This study uses C-SPARQL [1], which is a mainstream system for stream reasoning. C-SPARQL is a version of SPARQL 1.1[4] with an extension that allows the handling of stream data. With C-SPARQL, a user can monitor average values of a sensor by setting and then sliding a regular time interval at a designated time step. We use C-SPARQL for customer detection in a case study of PRINTEPS, which we will examine in Sect. 3.2

The link between ROS and C-SPARQL is obtained by implementing a Subscriber when the sensor data are published as an ROS Topic and by generating an RDF stream in the Subscriber based on the domain ontology from the Message-type object output from the Publisher. The use of the C-SPARQL library enables the RDF stream to detect an event based on the C-SPARQL query and to publish the detected result. Event-detection conditions can thus be described in the C-SPARQL query format and separately from the program with a high level of abstraction based on the domain ontology.

3 Actual Use for Teahouse Operations

In this section, we describe the integration of rules through events and image sensing in PRINTEPS as well as the actual application of PRINTEPS to teahouse customer-reception services.

The operations of a teahouse include receiving customers, guidance to tables, order taking, food preparation, serving, checkout, and expression of gratitude towards customers. This study focuses on customer-reception services using PRINTEPS.

3.1 Customer Reception at the Entrance Service

The "greeting to the customer" process detects an incoming customer by means of the Kinect sensor and orders the robot Pepper to give the customer a greeting based on the rule. Figure 1 shows the subprocesses that comprise the "greeting to the customer" process. Within this process, the "customer detection" process detects an incoming customer and then the "search of a customer-attending robot" module acquires the "robot IP," that is, the IP address of a robot that can offer the customer a reception service (Pepper's IP address). From the "robot

---

[4] http://www.w3.org/TR/sparql11-query/.

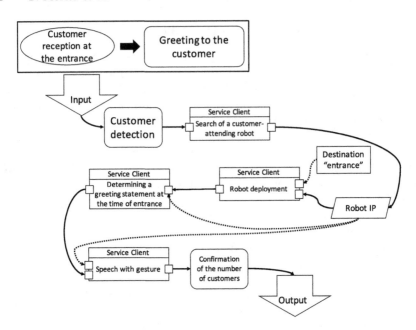

Fig. 1. Subprocesses and modules comprising the "Greeting to the customer" process.

IP" and the "destination" (which, in this case, is "the entrance" of the tea-house), PRINTEPS then retrieves the environment map and deploys Pepper to the entrance. The module entitled "determining a greeting statement at the time of entrance" determines, based on the rule, a greeting statement. The "speech with gesture" module orders Pepper to say a programmed greeting to the customer such as "Welcome to the shop!" Pepper is also programmed to ask the customer the question: "How many seats would you need?" to determine the number of people in the customer's party. The "confirmation of the number of customers" process acquires, through the speech and image recognition functions, the answer from the customer, and, if the number of party members determined by Pepper is different from the number given by the customer, the process corrects the initially determined number.

3.2   Customer Detection Process

Figure 2 shows the modules comprising the "customer detection" process shown in Fig. 1. The "entrance sensing" module, using Kinect SDK's person-detection library, continues to output ROS Topics indicating the number of people near the entrance and the distance between people and the Kinect sensor. The ROS Message on "Entrance" output from the "entrance sensing" module is shown in Fig. 3. The notation of Fig. 3 is nearly as same as the UML class diagram. The "Entrance.msg" data contains an ID, the number of customers (number_of_customers), and an array of the customer type (has_customers).

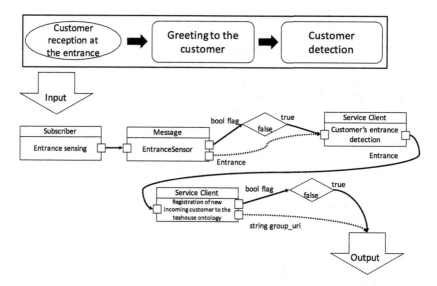

Fig. 2. Modules comprising the "customer detection" process.

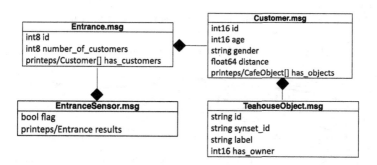

Fig. 3. ROS message on "entrance" output from the "entrance sensing" module.

The "Customer.msg" data contains the customer's ID, age, gender, distance between the customer's face and Kinect, and an array of the customer's objects (has_objects). The "TeahouseObject.msg" data contains the customer's ID, synset ID, and label that all correspond to the objects at WordNet[5]. The "customer's entrance detection" module subscribes the ROS Topic published by the "entrance sensing" module and, through the link with C-SPARQL, detects an incoming customer when the conditions of C-SPARQL shown in Fig. 4 are met. The module entitled "registration of new incoming customer to the teahouse ontology" then updates the ontology of the teahouse. In this module, based on the teahouse ontology shown in Fig. 5, an RDF stream for detecting a customer (Fig. 6) is generated from the Customer.msg's array of customer objects. In Fig. 5, ovals with solid line represent classes and ovals with dotted

---

[5] https://wordnet.princeton.edu/.

```
REGISTER QUERY CustomerDetectionQuery AS
PREFIX f: <http://larkc.eu/csparql/sparql/jena/ext#>
PREFIX rdf:<http://www.w3.org/1999/02/22-rdf-syntax-ns#>
PREFIX teahouse_class: <http://printeps.org/teahouse/class/>
PREFIX teahouse_property: <http://printeps.org/teahouse/property/>
SELECT ?s (COUNT(?s) AS ?cnt) (AVG(?distance1)
FROM STREAM <http://printeps.org/teahouse/entrance>
 [RANGE 3s STEP 1s]
WHERE {
 ?s rdf:type teahouse_class:Customer.
 ?s teahouse_property:positionAtTime ?ts1; teahouse_property:positionAtTime ?ts2.
 ?ts1 teahouse_property:distance ?distance1.
 ?ts2 teahouse_property:distance ?distance2.
 BIND(?distance2 - ?distance1 AS ?difference)
 FILTER(
 (f:timestamp(?s, teahouse_property:positionAtTime,?ts1)
  > f:timestamp(?s, teahouse_property:positionAtTime,?ts2))
 && 0.1 < ?difference)
 }
GROUP BY ?s
HAVING (AVG(?distance1) < ?distance_to_entrance && 1 < COUNT(?s))
```

Fig. 4. C-SPARQL query for detecting customers.

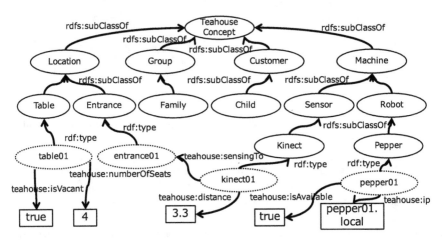

Fig. 5. Part of a teahouse ontology and its instance data.

line represent instances of several of the classes. This RDF stream is a set of RDF triples that contains: an instance of the customer's class (customer_1 shown in Fig. 6) generated for each ID of customers detected with Kinect; an instance generated as the positionAtTime property value showing the detected customer at a certain time (e.g., customer_1_1448426411267 in Fig. 6); and a printeps:distance

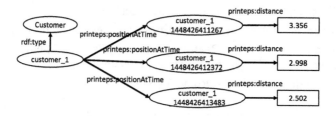

Fig. 6. RDF stream for detecting a customer.

property value showing a distance (m) between the detected customer's face at a certain time and Kinect. Finally, the module adds RDF triples containing information about a customer to the teahouse ontology.

Figure 4 shows the C-SPARQL query for detecting a customer. This query measures at every one-second interval the distance between the face of every person (detected within 3 s) and Kinect. The distances between the faces of the people detected within 3 s (ID: c1) and Kinect are chronologically shown as c1d1, c1d2, c1d3, and the values of c1d1-c1d2, c1d1-c1d3, and c1d2-c1d3, are computed in order to count the number of values greater than 0.1. The value 0.1 is determined based on the accident error of Kinect's depth sensor value. This measurement is used to avoid erroneously detecting someone who is in front of the teahouse but is not approaching it as a customer. A count exceeding 1 means that someone is approaching the teahouse (Kinect), which is a measurement used to avoid erroneously reacting with a customer who is leaving the teahouse. The average value of the distances between the faces of the people detected within 3 s and Kinect is also calculated, and if the average value is less than the ?distance_to_entrance (3.3 m), the customers are recognized as having entered the teahouse, and their IDs, the aforementioned count, and the average values are returned. The value of ?distance_to_entrance is obtained using a SPARQL query for measuring a distance between Kinect and an entrance by predefining, based on the teahouse ontology, the distance between the location in which Kinect is installed and the entrance of the teahouse. Currently, a distance between a person's face and Kinect is the only information that is used. However, if the sensor can obtain various attribute information from a person in the future, more complex customer detection based on such information (e.g., discerning a customer from a teahouse clerk based on clothing) will be realized.

3.3 Determining a Greeting Statement at the Time of Entrance Module

This section describes the most important module of the "greeting to the customer" process (Fig. 1). This is the module called "determining a greeting statement at the time of entrance". In PRINTEPS, the semantic web rule language (SWRL)[6] format rules that define the (business) rules governing the behaviors

---

[6] http://www.w3.org/Submission/SWRL/.

Entrance(?entrance), GreetingAtEntrance(?service),

Group(?group), Pepper(?pepper), hasGroup(?entrance, ?group),

robotPosition(?pepper, ?entrance), servedBy(?service, ?pepper),
servedTo(?service, ?group), numberOfCustomers(?group, 2)

-> greeting(?pepper, "Welcome. Two seats?")

Fig. 7. Example of greeting statement rule.

of a robot are automatically generated after the rules are defined in the ontology
editor Protégé[7] using the Manchester syntax. Figure 7 shows the rule that the
module applies when determining a greeting statement upon detecting two customers. In this study, we used the reasoning engine Pellet[8] to apply the greeting
properties and their values to the robot-class instance and to determine a greeting statement. The rule shown in Fig. 7 means that a group of two people is found
at the entrance; Pepper is positioned at the entrance; when Pepper provides a
service to the group, Pepper gives the statement: "Welcome! Two seats?"

In this particular case, the greeting statement rule is defined simply based
on the number of people in a group. If a teahouse owner wants a robot to give a
more detailed greeting statement based on a customer's age, gender, members of
the party (e.g., family members, girl or boyfriend), the owner must only revise
the rule described in the Manchester syntax defined there. However, under the
current setting, such revision requires the use of Protégé. Thus, for a teahouse
owner without knowledge of Protégé or ontology to update the rule on his or her
own is not easy. We plan to develop an editor that facilities the rule-updating
operation. Once it is developed, we will link it to the workflow editor so that the
end user can easily modify the rules.

4   Related Work

In this section, we introduce a related study: the RoboEarth Project [7]. We also
examine their differences based on our study.

RoboEarth [7] is a project focused on tasks related to autonomous robots,
such as robotics, knowledge processing, and environment sensing for understanding the actual environment. Knowledge processing for Robots (KnowRob) [6] is
one of RoboEarth's subprojects and provides several frameworks for knowledge
processing using an ontology, for example, and is closely related to this study.
Under KnowRob are structured multiple ontologies, which are used as knowledge expressions of tasks, robots, motions, and environment. KnowRob uses a
framework that allows its robots to conduct requested tasks by linking all the
knowledge contained in those ontologies. It aims to enable robots to exchange
information over the Web.

---

[7] http://protege.stanford.edu/.

[8] http://clarkparsia.com/pellet.

Regarding the integration of image sensing and knowledge processing, RoboEarth enables a robot to understand not only information it acquires through image sensing, but conceptual information as well, by adding Web Ontology Language (OWL)[9]-based knowledge expressions to the object models that are used for object recognition. Moreover, RoboEarth produces robots of different types to conduct tasks such as delivering a drink to a table by linking OWL-based knowledge expressions to environment maps of multiple types or linking to a combination of necessary programs for certain pre-defined tasks known as action recipes.

Two major differences exist between this study and RoboEarth.

The first is with respect to the targets of image sensing. In RoboEarth's image sensing, these are targets or obstacles necessary for a task for which no time-series changes must be considered. This means that the integration of image sensing and knowledge processing is achieved simply by adding knowledge expressions such as conceptual information to the object models. In addition, the targets of this study's image sensing are people involved in time-series changes (e.g., those who are going to enter a teahouse). Therefore, this study attempts to integrate dynamic information (events) acquired through image sensing with static information (i.e., business rules) by using C-SPARQL.

The second difference is in the correction frequency of workflow. The chief objective of RoboEarth is to enable robots to conduct specific tasks (e.g., delivering a drink to a table) in various environments based on a common workflow. Therefore, RoboEarth does not require that a domain expert directly and frequently change the workflow.

By contrast, this study requires that a domain expert (e.g., a teahouse owner) change the workflow of robot services frequently based on unique experiences of both robots and customers, the environments in which they operate, and reactions from customers. For that purpose, this study aims to build a platform that allows easy changes based on image sensing results by providing an SOA-based workflow editor that facilitates adjustments to robot services. These changes can also be accomplished by separating business rules from programs or by revising the knowledge processing mechanism.

5 Conclusion

In this study, we introduced an application of PRINTEPS for detecting events by using stream reasoning and ROS, and for integrating image sensing with knowledge processing. Based on this platform, we demonstrated that the behaviors of a robot in a robot cafe can be modified by changing the applicable rule sets.

Our remaining tasks include: applying stream reasoning to additional image sensing, thus enabling additional information to be used to further enrich services, and measuring the operator hours necessary to change a service in order to verify the ease of correction.

---

[9] https://www.w3.org/TR/owl-ref/.

Acknowledgement. We are grateful to Mr. Yusuke Nakayama of the Graduate School of Science and Technology of Keio University for his software-implementation for this study.

This study has been supported by "A Framework PRINTEPS to Develop Practical Artificial Intelligence" of the Core Research for Evolutional Science and Technology (CREST) of the Japan Science and Technology Agency (JST).

References

1. Barbieri, D.F., Braga, D., Ceri, S., Della Valle, E., Grossniklaus, M.: C-SPARQL: a continuous query language for RDF data streams. Int. J. Semant. Comput. **4**(1), 3–25 (2010)
2. Boyer, J., Mili, H.: Agile Business Rule Development: Process, Architecture, and JRules Examples. Springer, Heidelberg (2011)
3. Valle, E., Ceri, S., van Harmelen, F., Fensel, D.: It's a streaming world! reasoning upon rapidly changing information. IEEE Intell. Syst. **24**(6), 83–89 (2009)
4. Girshick, R., Donahue, J., Darrell, T., Malik, J.: Rich feature hierarchies for accurate object detection and semantic segmentation. In: 2014 IEEE Conference on Computer Vision and Pattern Recognition (CVPR), pp. 580–587 (2014)
5. Quigley, M., Conley, K., Gerkey, B.P., Faust, J., Foote, T., Leibs, J., Wheeler, R., Ng, A.Y.: Ros: an open-source robot operating system. In: ICRA Workshop on Open Source Software (2009)
6. Tenorth, M., Beetz, M.: KnowRob: a knowledge processing infrastructure for cognition-enabled robots. Int. J. Robot. Res. **32**(5), 566–590 (2013)
7. Waibel, M., Beetz, M., Civera, J., D'Andrea, R., Elfring, J., Galvez-Lopez, D., Haussermann, K., Janssen, R., Montiel, J.M.M., Perzylo, A., Schiessle, B., Tenorth, M., Zweigle, O., Molengraft, R.: Roboearth. IEEE Robot. Autom. Mag. **18**(2), 69–82 (2011)
8. Yamaguchi, T.: A platform printeps to develop practical intelligent applications. In: Adjunct Proceedings of the 2015 ACM International Joint Conference on Pervasive and Ubiquitous Computing and Proceedings of the 2015 ACM International Symposium on Wearable Computers, UbiComp/ISWC 2015 Adjunct, pp. 919–920. ACM (2015)

Quality of Thai to English
Machine Translation

Séamus Lyons[(✉)]

Department of Information Technology, International College,
Payap University, Chiang Mai, Thailand
seamus_l@payap.ac.th

Abstract. This paper presents the experimental results of several approaches to machine translation evaluation to determine the quality of Thai to English translation. We compare automatic metrics and human-based evaluation that includes error classification, reading comprehension and analysis from a professional translator. The research compares translation systems that are available to end users in Thailand to provide an understanding of the quality of translation in general use. Both the rate of 47.2 % error words per text and the BLEU score of 0.21 indicate the difficulty of Thai to English translation. Despite a high error rate for the translations, users were able to successfully answer about 60 % of the questions using the output of the machine translation systems in the reading comprehension tests.

Keywords: Machine translation · Machine translation evaluation · Error classification · Reading comprehension

1 Introduction

The task of machine translation (MT) evaluation is difficult as there is no definitive translation nevertheless it is necessary to determine the performance, differences and improvements of translation systems. Improvements seen in evaluation can inform and inspire researchers in future work. The purpose of this study is to ascertain both an indication of the level of quality of Thai to English translation, and to identify the significant problems for Thai to English translation. The study focuses on the evaluation of MT systems in general use in Thailand.

Quality is defined as the standard as measured against other things of a similar kind, or, the meeting of the requirements of the user. The purpose of translation use differs, but the general goal is a well-formed (fluent) translation expressing the same meaning (adequacy) as the source text. There are several evaluation approaches that measure the standard of translation systems such as using the presence and correction of errors. Minimal or no error is required when a publisher wishes to provide content in another language and a translation is considered 'publishable' with 5 % or less errors and 'editable' with about 15 % or less errors (Koehn 2013). Editable is when the translation makes computer assisted translation more productive than without it. Communication and assimilation, such as gist reading, are examples of when a lower error rate (e.g. 30 %) is tolerable. Research suggests Thai to English translation is within this band of

© Springer International Publishing Switzerland 2016
H. Ohwada and K. Yoshida (Eds.): PKAW 2016, LNAI 9806, pp. 261–270, 2016.
DOI: 10.1007/978-3-319-42706-5_20

lower quality. In a local survey (Lyons 2016), the participants gave a positive response for their satisfaction with Thai to English translation yet the few published evaluation scores for Thai MT are poor. The survey identified many methods to access translation but these only provided five different translation outputs. This research uses several techniques to evaluate the five MT systems that provide translation in current use.

There are numerous approaches to machine translation evaluation and we incorporate a combination of common and rare methods. BLEU (Papineni et al. 2002) is the predominant reported metric for machine translation research, largely due to its ease of use, official use in relevant workshops and correlation to human judgments. BLEU is an automatic evaluation metric based on the recall of N-grams, it measures the similarity of the output of MT systems to one or more reference texts. There are other automatic measurements but they are reported in conjunction with BLEU rather than as an alternative.

Our second approach is error classification which identifies, categorizes and quantifies the errors seen in the output of MT systems. We augment the error classification with the knowledge of a professional translator to give insight to the specific problems of Thai to English translation. Our final approach is reading comprehension, which is a rare form of evaluation because of logistic difficulties but it gives a natural way of assessing how the translations are understood by the end user.

The next section describes some relevant features of the Thai language and MT research in Thailand. Section 3 details the relevant approaches to MT evaluation and is followed by an explanation of the procedures for the evaluation methods. Section 5 includes an overview of the experimental results and the error analysis performed by the professional translator, this is followed by the conclusion.

2 Thai Machine Translation

The Thai language is a tonal language written in a script without spaces between words. In Thai each word has a complete, self-containing meaning or several meanings, dependent on the sentence word order and position of the word. The word may be modified or augmented by other words to indicate gender, noun number, tense or mood of a verb (Chimsuk and Auwatanamongkol 2009). For example, the Thai word หนังสือ (book) does not change when referring to 'books'. It is normal to state the amount of books with the appropriate classifier, เล่ม for books, underlined in Fig. 1a. The future tense is marked with จะ, used similarly as 'will' as in "I will read a book", and the past tense is understood by the meaning of other words, such as the use of เมื่อวานนี้ for 'yesterday' in Fig. 1b. In 2000, ParSit became the first Thai MT system available to the public (Sornlertlamvanich et al. 2000). It was designed to allow access to web pages written in English to Thai readers. Since then Statistical Machine Translation (SMT) has become the prominent approach to machine translation seen in the early Thai research of SMT (Netjinda et al. 2005) and phrase-based SMT research (Labutsri et al. 2009). A significant amount of Thai MT research has focused on the problem of segmenting text. The most successful approach uses Conditional Random Fields (CRF) seen in the comparison of six word segmentation systems in (Noyunsan et al. 2014). Thailand National Electronics and Computer Technology Center (NECTEC) have contributed much of the translation research in Thailand. Its continued research includes the implementation of a speech-to-speech translation system which is focused on the travel domain for ASEAN MT (Wutiwiwatchai 2015).

Fig. 1. Differences between (a) plurals and (b) verb tense in Thai and English.

MT systems have difficulty translating from Thai because of word and sentence segmentation, and the analysis and parsing of Thai grammar (Wutiwiwatchai et al. 2007). Speech translation research highlights segmentation as a major problem along with word reordering within a noun phrase and Named Entity detection (Wutiwiwatchai et al. 2009). A wide range of issues relating to unknown words have been researched in different institutes in Thailand. There are some positive results when these tasks are researched specifically for the Thai language such as in (Luekhong et al. 2013) for word alignment. Some research has continued but many areas were researched independently and future work did not follow.

3   Machine Translation Evaluation

There is little research that includes the evaluation of Thai to English translation. Error identification can be seen in a table of linguistic problems reported in (Supnithi et al. 2002), and the reasons for a poor BLEU score of 0.14 are discussed in (Nathalang et al. 2010). Other research of Thai to English translation reports a BLEU score of 0.23 (Slayden et al. 2010).

Error classification has become widely used to evaluate the strength and weaknesses of a translation system (Popović and Burchardt 2011). Vilar et al. (2006) describes a framework for a classification scheme of errors from a MT system that provided the basis of the five error classes used in (Popović and Ney 2007) consisting of inflectional error, reordering error, missing word, extra word and lexical choice error. The approach identifies the Word Error Rate (WER), a common metric for speech translation systems that calculates the minimum number of insertions, deletions and substitutions that have to be performed to convert the translation, called the hypothesis, into the reference text. Error rates stated in the introduction use the HTER metric where human post-editing calculates these errors. The WER and Position-independent word Error Rate (PER) are subsequently used in further error classification research (Popović and Ney 2011).

Comprehension evaluation was part of the Advanced Research Projects Agency (ARPA) methodology in 1991 to evaluate machine translations. It has not been used primarily because there are concerns about the difficulties in administering the tests. This problem was overcome with the use of Mechanical Turk to provide participants for the tests in (Callison-Birch 2009). Reading comprehension tests were also used at the 2008 NIST Open MT evaluation (Przybocki et al. 2008).

4 Experimental Methods

To perform reliable experiments on the MT systems and ensure unbiased results the source and reference texts needed to be original. A test set of 50 documents each containing about 75 words based on different domains was gathered from different sources, and paraphrased. The 50 articles were then translated by a professional translator into the Thai source text.

The experiments evaluate the output of five MT systems chosen because they provided the translations identified in use in a local survey of over 1700 participants (Lyons 2016). The five MT systems are Google Translate, Baidu, Naver MT used for 'Line', a communications app, the Microsoft Translator (Bing) and a MT system used by several providers such as freetranslation.com (SDL Language Weaver), and Babylon. For brevity these five MT systems are furthermore called 'Google', 'Bing', 'Baidu', 'Line' and 'Free'. A different professional translator analyzed the source (Thai) text and the output text of the five MT systems.

The reading comprehension tests involved a set of four questions for each of the 50 texts using guidelines taken from (Callison-Burch 2009). Questions were answerable in a few words, and did not ask about numbers or dates, or only require a yes/no answer. For each of the 50 texts, the output of one of the five MT systems, or the reference text, was randomly given with the set of four questions for each text resulting in a questionnaire of 200 questions. This approach led to six unique questionnaires each answered by two individuals, resulting in twelve participants, all Thai nationals.

The Hjerson tool for automatic error classification (Popović 2011) calculates the minimum number of insertions, deletions and substitutions that have to be performed to convert the translation, called the hypothesis, into the reference text. Position independent word Error Rate (PER) does not take the word order into account. The precision-based metric '*H*per' refers to the set of words that appear in the reference text but not in the hypothesis, and '*R*per' (recall-based) refers to the set of words that appear in the hypothesis but not in the reference text. To gain a metric for the word error rate independent of word order we use the average of the *H*per and *R*per metrics to form a combined error rate that is position-independent.

5 Experimental Results

The experiments resulted in low BLEU scores for the MT systems between 0.13 and 0.21 and high error rates in the region of 50 % of the words requiring insertion, deletion or substitution (see Table 1). These results are consistent with other reported

BLEU scores for Thai to English indicating the difficulty of the translation. There is not a significant difference between the systems in terms of approach as they all multilingual systems that incorporate phrase-based SMT.

Table 1. Average BLEU score and combined error rate for the five machine translation systems.

MT system	BLEU score	Combined error rate (%)
Google	0.209	47.2
Baidu	0.172	49.1
Bing	0.160	51.8
Free	0.146	53.0
Line	0.132	54.8

The outcome of the reading comprehension tests differed in that there were two distinct levels of quality. Two systems, Google and Baidu, scored just below 60 % of the questions correctly, whilst the other three systems scored just above the 50 % mark. This separation of the MT systems into two levels of quality was also seen in the error classification results that quantified errors into categories. Google and Baidu had similar error rates and in four of the five categories their error rate was lower than the other three systems.

Table 2 shows the error classification experiments using the Hjerson tool. The Reference position-independent error rates (Rper) are consistent over the five MT systems with less than four errors per text difference between the systems. The percentage of the errors to total words in the reference text is high for machine translation at about 50 %. We combine both the Reference and Hypothesis position-independent error rates to form one Combined Error Rate. The Combined Error Rate decreases as the BLEU scores increase, seen in Table 1, illustrating a strong correlation despite the different nature of the metrics.

Table 2. Reference (Rper) and Hypothesis (Hper) Position independent word error rates.

MT system	Rper	Reference text errors (%)	Hper	Hypothesis text errors (%)
Google	32.4	46.6	34.3	47.8
Baidu	33.4	48.2	36.2	50.0
Bing	34.5	49.7	41.6	53.9
Free	33.4	48.2	50.2	57.8
Line	36.2	52.5	44.2	57.0

The results of the reading comprehension tests for each MT system did not follow the same pattern as the BLEU scores and error rates (see Table 3). The scores in the reading comprehension test showed two distinct levels of quality with both Baidu (59.75 %) and Google (59 %) scoring about 60 % of the questions answered correctly, and the three other systems scoring just above 50 % (Bing, 53 %, Free, 51 %, Line, 50.75 %). There was a considerable amount of MT outputs that contained the information required to answer questions but the participant did not fully understand the text or the question. An alternative approach for extensive testing could use a range of difficulty levels for the questions. This is common in language assessment with the IELTS test for English proficiency having nine levels of assessment, and levels in the Defense Language Proficiency Test (DLPT), a measure of effectiveness for evaluating foreign language proficiency, used for reading comprehension tests in (Jones et al. 2005). In the comprehension tests, the MT systems scores should perhaps be compared to the scores obtained using the reference text to give an accurate reflection of their performance.

Table 3. Results of the reading comprehension tests for the machine translation systems and reference text.

	Correct answers	Percentage (%)
Reference text	326	81.5
Baidu	239	59.75
Google	236	59
Free	212	53
Bing	204	51
Line	203	50.75

Despite the various levels of English competence the participants successfully answered 81.5 % of the questions when using the reference text. The best MT system performance (59.75 %) is 73.3 % of the reference text score (81.5 %). This gives a positive reflection on the capability of the MT system to provide a translation of a lower level of quality such as gist reading or communication. It is reported that SMT systems are trained to optimize their BLEU score and it is interesting to see a different system, Baidu, match Google in an alternative evaluation method. These results are not conclusive but they do question the assumption that there is a significant difference between Google and other MT systems.

The error classification experiments that categorized the errors are based on the subset of 100 translations from the five MT systems used for the error analysis by the professional translator. Word meaning ambiguity is the most significant problem for MT systems. In previous studies, 'wrong lexical choices' were the most frequent errors found (Avramidis et al. 2012). This category was found to be the most problematic, both in the error analysis and in the error classification results, seen in Table 4. The 'incorrect lexical choice' category accounted for over half the errors (54.8 %). Perhaps surprisingly the error classification tool found the 'missing words' category contained only 5.3 % of the errors.

The error analysis performed by the translator identified some explanations for the errors identified in the error classification. Examples can be seen in Fig. 2, in the translation of the text meaning "the body of an Australian man". Ambiguity in word meaning is seen in translation #2 where the Thai word 'ร่าง' has several meanings including 'body' and 'draft'. Segmentation is also a prominent issue seen in translation #4 where the MT system fails to identify which words should be grouped to ascertain

Table 4. Error classification results

Error category	Baidu	Free	Google	Bing	Line	Total	Total (%)
Inflectional errors	58	61	56	64	42	281	5.8
Incorrect word order	168	197	174	192	183	914	19.0
Missing words	62	22	62	59	51	256	5.3
Extra words	96	252	80	132	163	723	15.0
Incorrect lexical choice	488	571	467	534	579	2640	54.8

the correct meaning. This problem is also seen with the frequent use of compound words. For example, the common words 'น้ำ' and 'เงิน' meaning 'water' and 'money' are joined to mean 'navy blue' in 'น้ำเงิน'. Errors in word order are seen in both translations #2 and #5 with the later finding difficulty segmenting the text correctly. Finally, the Thai text does not indicate singular or plural resulting in translations #1 and #5 pluralizing 'man' to 'people'.

Source Text	ร่างของชายชาวออสเตรเลียคนหนึ่ง
Reference Text	The body of an Australian man
Translation #1	The body of the Australian people
Translation #2	The draft of the man Australian
Translation #3	The body of an Australian man
Translation #4	The body of a man who believes that Australia …
Translation #5	a man's Australian people

Fig. 2. Example of the output of MT systems for a simple clause

The omission of a pronoun is not unusual in spoken Thai and causes problems when the object noun, or a pronoun, is omitted in text. In the example in Fig. 3a the object 'the bear' is not repeated or replaced by a pronoun. This results in the translation error where the words 'I know' replace the correct translation of 'the bear knew'. The meaning of the pronoun can also depend on the context such as the Thai term 'เอง' that translates to self, himself, oneself, yourself, or myself.

In Thai the correct meaning of a combination of words is altered if a word is missing or mistranslated. This is also seen in English with negation, such as 'have' and 'have not'. In Fig. 3b the underlined text means "was trapped for some time", seen in the reference text. But the meaning is dependent on the Thai word 'ใกล้', meaning 'near', also underlined in the Thai text and found at the end of the text followed by the Thai symbol 'ๆ' which repeats the preceding word for emphasis. There is a space in the Thai text between the original text and 'ใกล้' creating 'long distance' difficulties for the MT system. This results in the incorrect translation 'is the time' in the example translation. There were also many unexplainable errors in the translations. For example, "in the month of November" was mistranslated as "in the month of 6,800". Pure statistical MT systems are prone to some anomalies leading to some researchers suggesting additional linguistic analysis is required.

Source Text	เรื่องนี้พอทราบหมีก็ตื่นเต้นเลยเพราะทราบดีว่าเหล่าหมาป่านั้นจะต้องเผชิญกับอะไรบ้าง
Reference Text	It was greeted with hysteria as the bear knew what the wolves were going to suffer.
Translation	This bear was excited because **I know** enough to know that these wolves are faced with something.

(a)

Source Text	หมาป่าตกลงไปในบ่อน้ำและติดอยู่ในนั้นเป็นระยะเวลานาน
	มันได้ยินเสียงแพะเดินเข้ามาใกล้ๆ
Reference Text	A fox fell down a well and <u>was trapped for some time</u>. He heard a goat approaching …
Translation	the fox fall into the water and <u>trapped in there, is the time,</u> it heard the goat walked in

(b)

Fig. 3. (a) Example of the problem of a missing noun or pronoun, and (b) meaning alteration by a sequence ending word.

6 Conclusion

The level of quality of Thai to English translation in use is the standard as measured against other MT systems seen in the BLEU score of 0.21 and an error rate of 47.2 %. The comprehension tests gave a better indication of how much these errors effect the ability of the system to meet the requirements of users. The level of quality, stated less formally, was the ability of a user to answer six out of ten questions correctly using MT output, as opposed to eight out of ten when using the reference text. Comprehension tests were dismissed because of logistic difficulty and expense, yet it is common in other areas of language evaluation to use the ability to answer questions with levels of difficulty to indicate a level of capability.

Some Thai translation issues such as word order will be largely resolved with additional resources for SMT such as the availability of bilingual corpora if made publically available from ASEAN. Other problems such as managing unknown words require further research. Segmentation is problematic for multilingual MT systems and illustrates the need for research focused solely on the translation of the Thai language. The use of several evaluation techniques giving insight into the ability of users to perform required tasks, in conjunction with work focused on Thai translation, could motivate researchers to provide an improved translation service to end users.

References

Avramidis, E., Burchardt, A., Federmann, C., Popović, M., Tscherwinka, C., Vilar, D.: Involving language professionals in the evaluation of machine translation. In: LREC, pp. 1127–1130 (2012)

Callison-Burch, C.: Fast, cheap, and creative: evaluating translation quality using amazon's mechanical turk. In: Proceedings of the 2009 Conference on Empirical Methods in Natural Language Processing, vol. 1, pp. 286–295. Association for Computational Linguistics (2009)

Chimsuk, T., Auwatanamongkol, S.: A Thai to English machine translation system using Thai LFG tree structure as interlingua. World Academy of Science, Engineering and Technology, pp. 690–695 (2009)

Koehn, P.: Open problems in machine translation (2013). https://www.youtube.com/watch?v= 6UVgFjJeFGY

Jones, D., Shen, W., Granoien, N., Herzog, M., Weinstein, C.: Measuring translation quality by testing English speakers with a new defense language proficiency test for Arabic. Massachusetts Institute of Technology, Lexington Lincoln Lab (2005)

Labutsri, N., Chamchong, R., Booth, R., Rodtook, A.: English syntactic reordering for English-Thai phrase-based statistical machine translation. In: Proceedings of the 6th International Joint Conference on Computer Science and Software Engineering (JCSSE 2009) (2009)

Lyons, S.: A survey of the use of mobile technology and translation tools by students at secondary school in Thailand. Payap Univ. J. **26**(1) (2016)

Luekhong, P., Ruangrajitpakorn, T., Supnithi, T., Sukhahuta, R.: Pooja: similarity-based bilingual word alignment framework for SMT. In: Proceedings of the 10th International Symposium on Natural Language Processing, Phuket, Thailand (2013)

Netjinda, N., Facundes, N., Sirinaovakul, B.: Toward statistical machine translation for Thai and English. In: International Symposium On Digital Libraries, Albuquerque, New Mexico, USA, 27–28 October 2009 (2005)

Noyunsan, C., Poltree, C.H.S., Saikeaw, K.R.: A multi-aspect comparison and evaluation on Thai word segmentation programs. In: JIST (Workshops & Posters) 2014, pp. 132–135 (2014)

Nathalang, S., Porkeaw, P., Supnithi, T.: Don't use big words with me: an evaluation of English-Thai statistical-based machine translation. In: Proceedings of the International Symposium on Using Corpora in Contrastive and Translation Studies (UCCTS2010) (2010)

Papineni, P., Roukos, S., Ward, T., Zhu, W.-J.: BLEU: a method of automatic evaluation of machine translation. In: Proceedings of the 40th Annual Meeting on Association for Computational Linguistics, pp. 311–318. Association for Computational Linguistics (2002)

Popović, M.: Hjerson: an open source tool for automatic error classification of machine translation output. Prague Bull. Math. Linguist. **96**, 59–67 (2011)

Popović, M., Burchardt, A.: From human to automatic error classification for machine translation output. In: 15th International Conference of the European Association for Machine Translation (EAMT 2011) (2011)

Popović, M., Ney, H.: Word error rates: decomposition over POS classes and applications for error analysis. In: Proceedings of the Second Workshop on Statistical Machine Translation, pp. 48–55. Association for Computational Linguistics (2007)

Popović, M., Ney, H.: Towards automatic error analysis of machine translation output. Comput. Linguist. **37**(4), 657–688 (2011)

Przybocki, M.A., Peterson, K., Bronsart, S.: Translation adequacy and preference evaluation tool (TAP-ET). In: LREC, vol. 2008, p. 6 (2008)

Slayden, G., Hwang, M.Y., Schwartz, L.: Thai sentence-breaking for large-scale SMT. In: 23rd International Conference on Computational Linguistics, p. 8 (2010)

Sornlertlamvanich, V., Charoenpornsawat, P., Boriboon, M., Boonmana, L.: ParSit: English-Thai machine translation services on internet. In: 12th Annual Conference, ECTI and New Economy. National Electronics and Computer Technology Center, Bangkok (2000)

Supnithi, T., Sornlertlamvanich, V., Charoenporn, T.: A cross system machine translation. In: Proceedings of the 2002 COLING Workshop on Machine Translation in Asia (COLING-MTIA 2002), vol. 16 (2002)

Vilar, D., Xu, J., d'Haro, L.F., Ney, H.: Error analysis of statistical machine translation output. In: Proceedings of LREC, pp. 697–702 (2006)

Wutiwiwatchai, C.: Language and speech translation activities in Thailand. ASEAN-NICT Round Table – Feb 2015 (2015)

Wutiwiwatchai, C., Supnithi, T., Kosawat, K.: Speech-to-speech translation activities in Thailand. In: Workshop on Technologies and Corpora for Asia-Pacific Speech Translation (TCAST), p. 7 (2007)

Wutiwiwatchai, C., Supnithi, T., Porkaew, P., Thatphithakkul, N.: Improvement issues in English-Thai speech translation. In: Proceedings of TCAST Workshop 2009 (2009)

Stable Matching in Structured Networks

Ying Ling[1], Tao Wan[2(✉)], and Zengchang Qin[1(✉)]

[1] Intelligent Computing and Machine Learning Lab, School of ASEE,
Beihang University, Beijing 100191, China
zcqin@buaa.edu.cn
[2] School of Biological Science and Medical Engineering,
Beihang University, Beijing 100191, China
taowan@buaa.edu.cn

Abstract. Stable matching studies how to pair members of two sets with the objective to achieve a matching that satisfies all participating agents based on their preferences. In this research, we consider the case of matching in a social network where agents are not fully connected. We propose the concept of *D-neighbourhood* associated with *connective costs* to investigate the matching quality in four types of well-used networks. A matching algorithm is proposed based on the classical Gale-Shapley algorithm under constraints of network topology. Through experimental studies, we find that the matching outcomes in scale-free networks yield the best average utility with least connective costs comparing to other structured networks. This research provides insights for understanding matching behavior in social networks like marriage, trade, partnership, online social and job search.

Keywords: Stable matching · Structured networks · D-neighbourhood · Connective cost

1 Introduction

Stable matching can be best explained by the example of marriage and thus also known as the *stable marriage problem* (SMP). It aims to find a stable matching between two equally sized sets of elements given an ordering of preferences for each element. Two sets can be illustrated as an equal number n of men and women, in which every man ranks the n women according to how desirable of each is to him, without ties. Similarly, every woman ranks the n men based on their willingness (Gale and Sotomayor 1985). Ideally, a perfect match would pair every man with the woman he likes best and vice versa. However, preferences expressed by men and women rarely allow for a perfect match. But we can go for a *stable match*, such that there is no man and woman that both like each other better than their current partners. In order to obtain a stable match, we can start with random matching, exchange the unstable pairs by switching their partners until no pairs have motivation to change. Such a solution is known as the Gale-Shapley (G-S) algorithm (Gale 2013).

© Springer International Publishing Switzerland 2016
H. Ohwada and K. Yoshida (Eds.): PKAW 2016, LNAI 9806, pp. 271–280, 2016.
DOI: 10.1007/978-3-319-42706-5_21

The classical Gale-Shapley[1] algorithm assumes that all the information is known to public as complete information. Some works have reported to study the stable matching with incomplete information (Liu et al. 2014), such incompleteness may have a significant impact on the matching results. In social and economic interactions, an agent's well being depends on his or her own actions as well as on the actions taken by his or her neighbours. Such neighbouring relations can form a network and its structure decides the direct interaction. In recent years, the games played in networks have been studied extensively (Shoham 2008). A general framework for the study of games in such an incomplete-information setup has been developed in (Jackson and Watts 2002) and (Jackson 2005). Some related research has even developed into an independent area known as *Algorithmic Game Theory* (Nisan et al. 2007). However, few work has been done to study stable matching in networks. In this paper, we assume that the acquaintance between agents can be modeled by a network, a fully connected network indicates the ideal complete information. We are interested in stable two-sided matching in networks of different structures and the cost of matching within a network. We hope to understand how the patterns of social connections shape the choices that individuals make in matching.

The remainder of the paper is structured as follows. In Sect. 2, we introduce the basics of graph theory and four classical network structures that we will test later. In Sect. 3, we define D-neighbourhood for an agent in network and the cost of matching. A simple matching algorithm for network is proposed based on the classical G-S algorithm. Experimental results are given and analyzed in Sect. 4. In Sect. 5, we conclude this research and discuss the possible future work.

2   Network Structures

A social network is a social structure made of a set of agents and a set of the dynamic ties between them. In this paper, we mainly consider the following four types of well-studied networks: scale-free networks (Barabási-Albert model) (Barabási and Albert 1999), random networks (Erdös-Rényi model) (Erdös and Rényi 1959), small world networks (Watts-Strogatz model) (Strogatz 2001) and nearest-neighbor coupled network (NCN model). The reason for choosing these four structures is because they are representative social networks in other studies including (Li and Qin 2014) and (Li et al. 2013). In graph theory, a network can be viewed as a graph $G = (V, E)$, which is composed of a set of nodes V and edges E. Node number $N = |V|$, where $|.|$ represents the cardinality of a set, and the number of edges is $M = |E|$.

Barabási-Albert (BA) model is a typical scale-free network generation algorithm using a *preferential attachment*[2] mechanism. It reflects how normal social networks are formed, particularly online (Kitsak et al. 2007). The network is

---

[1] With Alvin E. Roth, Shapley won the 2012 Nobel Memorial Prize in Economic Sciences for the theory of stable allocations and the practice of market design.

[2] Preferential attachment can be regarded as a positive feedback in a network, more connected a node is, the more likely it is to receive new links.

Table 1. Average path length (APL) of four classical network models.

Network Model	NCN	ER	WS	BA
APL	$\overline{d} \propto N$	$\overline{d} \approx \frac{lnN}{ln\overline{k}}$	$\overline{d} = \frac{\sum_{i>j} d(i,j)}{N(N-1)/2}$	$\overline{d} \propto \frac{\log N}{\log \log N}$

seeded with two random links. Each link is given a weight equal to the degree of the target node it connects to, and a link is chosen in proportion to these weights. In the Erdös-Rényi model, $ER(N, p)$ is a graph constructed by connecting N nodes randomly with probability p independently from every other edge (Gomez-Gardenes and Moreno 2006). As a transition from the completely regular network to the completely random network, the introduction of a little randomness into regular network can generate a network with small world characteristics, known as Watts-Strogatz (WS) small-world network model (Latora and Marchiori 2001). Nearest neighbor-coupled network $NCN(N, k)$ of periodic boundary condition forms a ring of N vertex, where each node and its neighbors around are connected, k is an even number.

The topology of the network decides the dynamics of the network, two parameters characterizing complex network topology are well used (Wang and Jiang 2011): degree distribution, the average path length (APL). The degree k_i of the node i refers to the number of edges connected to the node i. The *average degree* of all nodes in a network is denoted by $\overline{k}$: $\overline{k} = \left(\sum_{i=1}^{N} k_i \right) / N$. Degree distribution is the probability distribution of node degrees over the whole network. Distance between two nodes i and j, $d(i, j)$ is defined as the number of edges in the shortest path connecting the two nodes using Dijkstra's algorithm (Dijkstra 1959), also referred to the Dijkstra distance. APL of the network is defined by Dijkstra distance: $\overline{d} = \frac{\sum_{i>j} d(i,j)}{N(N-1)/2}$. The equations of calculating average path length $(\overline{d})$ of four network models are shown in Table 1 (Wang and Jiang 2011).

3 Matching Model in Networks

We will be concentrating on *two-sided matching* markets (Roth and Sotomayor 2006) in this paper. Two-sided refers to the fact that agents in such games belong to one of two disjoint sets. In the real-world, regional limitation and attenuation of information flow help us to develop neighbourhoods, it also implicitly divided agents into groups and it is always costly to interact with agents far away. This fact inspires this study in order to understand how the changes in network structure will reshape the matching outcomes.

3.1 D-neighbourhood

Definition 1 *(D-neighbourhood). D-neighbourhood defines the nodes within the maximum permissible contact range. Given a maximum depth (D) for agent i,* *agent j satisfies that the Dijkstra distance $d(i,j)$ is less than D could achieve mutual acquaintance.*

$$\Delta(i,j|D) = \begin{cases} 1 & d(i,j) \leq D \\ 0 & d(i,j) > D \end{cases} \tag{1}$$

$\sum_j \Delta(i,j|D)$ calculates how many nodes are with distance D to the node i in a given network. Given $d(i,j) = l$ ($0 \leq l \leq D$), the least path is a sequence from the starting node i (for mathematical convenience, it can be denoted by κ^0) to the end node j (denoted by κ^l) through some specific intermediate nodes, or formally:

$$\langle i(\kappa^0), \kappa^1, \kappa^2 .... \kappa^{l-1}, j(\kappa^l) \rangle \tag{2}$$

Figure 1 shows an example of a network with the starting node i (in green) and the end node j. The nodes with distance 1 from node i are in blue and the nodes with distance 2 are in red. The shortest path between nodes i and j is $\langle i(\kappa^0), \kappa_2^1, j(\kappa_3^2) \rangle$ with $d(i,j) = 2$, but not the path of $\langle i(\kappa^0), \kappa_3^1, \kappa_2^1, j(\kappa_3^2) \rangle$ or other alternative paths. Where κ_s^t represents the sth node in the set of nodes with distance t to the starting node. In order to find the least length path in whole network, we need to choose κ_2^1 from five blue nodes within node i's distance 1, the probability to choose κ_2^1 is $P(\kappa_2^1) = 1/5$. The next node has to be chosen from 3 red nodes with distance 2 to the starting node i (κ^0), but only two of them have the distance 1 to κ_2^1. So $P(\kappa_3^2) = 1/2$, or formally, the probability of a node appearing in the least length path can be calculated by:

$$P(\kappa^s) = \frac{1}{\sum_t \Delta(\kappa^{s-1}, \kappa_t^s|1)} \tag{3}$$

$$s.t. : P(\kappa^0) = 1 \tag{4}$$

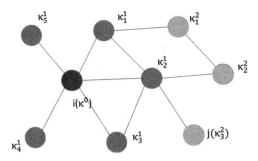

Fig. 1. An example of least length path from node i to j in a given network. The nodes are colored based on the distance to the starting node i (κ^0). (Color figure online)

Definition 2 *(Connective Cost). Connective cost of a matched pair $c_{i,j}$ measures the cost for agent i to know j through the intermediated nodes (Eq. (2)) between them.*

$$c_{i,j} = \prod_{d=1}^{l} \log\left(\frac{1}{P(\kappa^d)}\right) * \exp(d) \tag{5}$$

The connective cost is constructed by considering two factors: $c_{i,j} \propto \frac{1}{P(\kappa^d)}$, the lower probability a node has, the larger cost for it to get connected. $c_{i,j} \propto \exp(d)$ implies that, the increase of cost grows exponentially with the increase of depth. The reason of using logarithm is to re-scale the cost (which increases exponentially) when the network gets really large. The average connective cost between all matching pairs in a network with N nodes ($N/2$ probable matched pairs) is:

$$C = \sum_{ij} \frac{c_{i,j}}{N/2} \tag{6}$$

Definition 3 *(Network Connectivity). Network connectivity ϕ of a network refers to the proportion of the number of paths whose lengths are less than the maximum depth (D) to the number of all possible paths in the network.*

$$\phi = \frac{count(d \leq D)}{N(N-1)/2} \tag{7}$$

Connectivity for the classical G-S algorithm is considered as $\phi_{GS} = 1$. Actually, APL in each social network forms the difference in connectivity at the start. The distribution of the shortest path length of the random network (ER) obeys Poisson distribution: $P(X = d) = \frac{\lambda^d e^{-\lambda}}{d!} (d = 1, 2, 3...)$, where λ is APL. Then, the connectivity can be formulated by D, d and λ: $\phi = \int_0^D \frac{\lambda^d e^{-\lambda}}{d!} \mathrm{d}(d)$. Through theoretical derivation, APL of random network (ER) is negatively correlated with connectivity (i.e., $\lambda \uparrow \rightarrow \phi \downarrow$). From this, APL is basic and intrinsic characteristics of a network. Relationship of APL and connectivity in the four types of models will be tested in experimental studies.

3.2 Matching Model

There is a large collection of literatures on the matching models for markets with two-sided heterogeneity, such as the matching problems of students and schools, husbands to wives, and workers to firms (Roth and Sotomayor 2006) (Moldovanu 1992). Typical assumption of complete information makes the analysis tractable but stringent. Let us reconsider the problem in the marriage setting: there is a finite set of women, I, with an individual woman is denoted by $i \in I$. There is also a finite set of men, J, with an individual man $j \in J$. A matching pair function $\gamma : I \rightarrow J$, $\gamma(\cdot)$ is a bidirectional symmetrical mapping between I and J. If woman i's preference to man j is denoted by $R_{i,j}^w$, and man j's preference over woman i is $R_{i,j}^m$. Women or men can only give preferences of the ones within his (her) D-neighbourhood, it is an incomplete preference list comparing to the classical stable marriage problem. The satisfaction of an agent in matching can be defined as the following.

Definition 4 *(Satisfaction of Agent). Satisfaction of an agent measures how well his (her) preference list is meet in matching. The satisfaction for the woman $i(i \in I)$ is*

$$s_i^w = n_w - R_{i,\gamma(i)}^w \quad for \ i \in I \tag{8}$$

$$s.t. : \ \Delta(i, \gamma(i)|D) = 1 \tag{9}$$

The satisfaction for a man $j(j \in J)$ is

$$s_j^m = n_m - R_{\gamma^{-1}(j),j}^m \quad for \ j \in J \tag{10}$$

$$s.t. : \ \Delta(\gamma^{-1}(j), j|D) = 1 \tag{11}$$

where $n = N/2$ is the number of men (or women).

To avoid trivial cases, unmatched agents are assigned with zero satisfaction: $s_{i\varnothing} = s_{\varnothing j} = 0$. We then define a utility function of a matching pair through the satisfaction measure: $u_{i,j} = \frac{10(s_i^w + s_j^m)}{N}$. And the average utility of the matching is defined by: $U = \frac{\sum_{ij} u_{i,j}}{N/2}$. We consider one-to-one matching (i.e. no polygamy), with incomplete preference lists. The pseudo-code is shown in Algorithm 1.

Algorithm 1. Stable Matching Algorithm in Structured Networks

---

Inputs: Network G, D and preference lists R^w and R^m
Outputs: Matching outcomes ($\gamma : I \rightarrow J$)
while(for every man $j \in J$, if j is free)
   $i \leftarrow j$'s top woman in his preference list he never proposed to before
   if i is free
     (i, j) become a match
   else i have matched with j'
     if i prefers j' to j
       j stays free and propose to the next ranking woman i'
         if i' is beyond D-neighbourhood of j
           j stays free in this round
     else i prefers j to j'
       (i, j) become a match
       j' becomes free

A stable matching in a social network means there is no woman-man combination (i, j) such that $u_{i,j} > u_{i,\gamma(i)}$ and $u_{j,i} > u_{j,\gamma^{-1}(j)}$ for all (i, j) satisfying $\Delta(i, j|D) = 1$. It is stable if there is no unmatched man-woman pair that could increase both their utility by matching each other within their D-neighbourhoods. Comparing to the stability with complete information, our model may end with some men and women unmatched as they are not acquainted to each other. In the following experiments, we ensure every network is implemented under the same conditions with N and k are fixed.

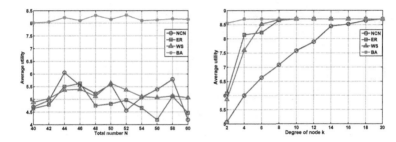

Fig. 2. Left-hand side: average utility of four networks with increasing number of agents with $k = 2$. Right-hand side: average utility of four networks with increasing node degrees with $N = 100$.

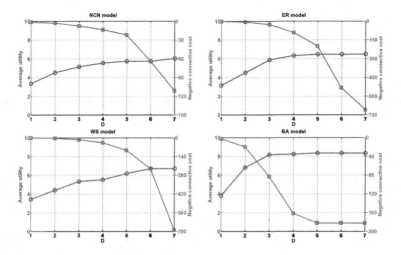

Fig. 3. Trade-off between average utility and average connective cost in four network models. Results are obtained by setting $N = 100, k = 2$. (Color figure online)

4   Experimental Studies

As we have discussed in previous sections that network topology may influence the matching outcomes significantly. In this section, we conduct matching experiments in small-scale networks with different structures. In each round, each agent is assigned with a preference list over all potential partners: $R \in [1, 10]$. While these networks are considerably smaller than the real networks, we set $D = 3$ as the maximum depth between any recognizable participants. Four types of networks introduced in Sect. 2 (NCN, ER, WS and BA) are tested and the results of average utility against on total numbers of agents is shown in the left-hand side of Fig. 2. The average utility is relatively stable given different number of agents, but BA is obviously with much higher average utility (indicates better matching) comparing to the other 3 models. The right-hand side figure shows the relation of average utility and node degrees. As we can see from the figure,

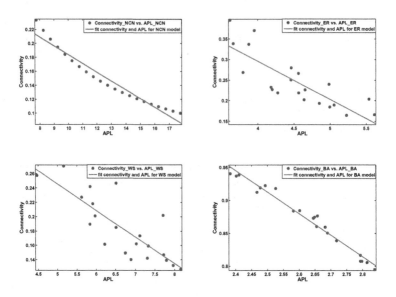

Fig. 4. Scatter plots of connectivity and average path length (APL) of four networks with $N = 100$.

given N is fixed, the larger k yields better utility of matching outcome. When k becomes large enough, all networks become fully connected and it converges to the situation of complete information as well as the average utility. Over all, BA still has the superior performance comparing to other 3 models.

As we have discussed in previous sections, the connective cost for knowing someone through others within your D-neighbourhood is calculated by Eq. (6). There is a trade-off between average utility and average connective cost defined based on the radius of one's D-neighbourhood. In Fig. 3, for each network model, we depict the average utility by circled blue curves and *negative* connective cost by squared red curves in double coordinates. Utility is increased, the connective cost is also increased (negative cost decreases) significantly. For each network model, we can focus on the intersection between the utility and cost curves. Comparing to other networks, the BA model has the most desired properties that the utility can reach 8 with cost of 80 at the depth of 3. For other three networks, the best utility values are less than 7. Though the NCN model has lower cost, the increase of utility is slow. Most importantly, such performance comparisons are conducted among four network models, even the definition of connective cost is modified with different parameters. The superiority of the BA model still holds.

In order to enlarge one's D-neighbourhood, we can either increase breadth (node degrees) k or the maximum depth D. When these two parameters are big enough, the network can be fully connected and becomes the classical stable matching problem with complete information. Table 2 gives the relations between connectivity (ϕ) and average utility (U) in four networks under different

Table 2. The connectivity and average utility of four classical network models with $k = 2$.

	NCN		ER		WS		BA	
	ϕ	U	ϕ	U	ϕ	U	ϕ	U
N=20	0.350	4.600	0.330	4.250	0.405	6.100	0.975	8.000
N=40	0.175	4.613	0.164	3.625	0.219	5.175	0.914	8.300
N=60	0.117	4.039	0.248	5.689	0.154	4.406	0.910	8.322
N=80	0.088	4.978	0.166	5.191	0.109	5.069	0.748	8.325
N=100	0.070	4.808	0.145	5.392	0.087	4.982	0.738	8.330

population sizes. No matter in which networks, the larger connectivity always indicates larger average utility of matched agents.

We have discussed that enlarged D-neighbourhood can make matching more efficiently. To give more quantitative and direct analysis, the scatter plots of average path length (APL) and connectivity of all four network models are shown in Fig. 4. Connectivity is negatively correlated to APL of a network. It means that more connected a network is, the shorter ALP we have. The connectivity of BA is much bigger than other networks which means that the agents have more opportunities to know other agents given the same radius of D-neighbourhood. It has shorter APL also means the less connective cost in matching. It gives a clue why BA may yield the best matched utility with less connective costs.

5   Conclusion

In this paper, we propose a stable matching algorithm by considering incomplete information in structured networks, where agents in both sides are not fully connected to each other. In reality, it can be interpreted as a marriage problem with limited acquaintances within a community. We considered four types of well-used networks and defined the *D-neighbourhood* and *connective cost* to imitate a real social network. Through simulated matching experiments, we found that the BA model has the most desired *average utility* with less connective costs. Thus it is the most efficient network among the four types of well-used networks in our experiments. We also investigated the relations among the network connectivity, average path length and average utility of matching. Empirical studies indicates that the reason BA is superior to others is mainly because it has a better connectivity allowing more matching opportunities for unmatched agents. Given the proposed matching algorithm, scale-free network has the best efficiency with low cost in matching. We will consider the case of one-to-many (school-student or job search) matching in structured networks as our future work.

Acknowledgement. This work is supported by the National Science Foundation of China Nos. 61305047 and 61401012.

References

Barabási, A.L., Albert, R.: Emergence of scaling in random networks. Science **286**(5439), 509–512 (1999)

Dijkstra, E.W.: A note on two problems in connection with graphs. Numerische Math. **1**(1), 269–271 (1959)

Erdös, P., Rényi, A.: On random graphs. Publicationes Math. **6**(4), 290–297 (1959)

Gale, D.: College admissions and the stability of marriage. Am. Math. Mon. **69**(5), 9–15 (2013)

Gale, D., Sotomayor, M.: Some remarks on the stable matching problem. Discrete Appl. Math. **11**(3), 223–232 (1985)

Gomez-Gardenes, J., Moreno, Y.: From scale-free to Erdos-Renyi networks. Phys. Rev. E **73**(5), 056124 (2006)

Jackson, M.O.: Allocation rules for network games. Games Econ. Behav. **51**(1), 128–154 (2005)

Jackson, M.O., Watts, A.: The evolution of social and economic networks. J. Econ. Theory **106**(2), 265–295 (2002)

Kitsak, M., et al.: Betweenness centrality of fractal, nonfractal scale-free model networks, tests on real networks. Phys. Rev. E **75**, 056115 (2007)

Latora, V., Marchiori, M.: Efficient behavior of small-world networks. Phys. Rev. Lett. **87**(19), 198701 (2001)

Li, Z., Qin, Z.: Impact of social network structure on social welfare and inequality. In: Pedrycz, W., Chen, S.-M. (eds.) Social Networks: A Framework of Computational Intelligence. SCI, vol. 526, pp. 123–144. Springer, Switzerland (2014)

Li, Z., Chang, Y.-H., Maheswaran, R.: Graph formation effects on social welfare and inequality in a networked resource game. In: Greenberg, A.M., Kennedy, W.G., Bos, N.D. (eds.) SBP 2013. LNCS, vol. 7812, pp. 221–230. Springer, Heidelberg (2013)

Liu, Q., et al.: Stable matching with incomplete information. Econometrica **82**(2), 541–587 (2014)

Moldovanu, B.: Two-sided matching-A study in game-theoretic modeling and analysis (Book Review). J. Econ. (1992)

Nisan, N., et al.: Algorithmic Game Theory. Cambridge University Press, Cambridge (2007)

Roth, A.E., Sotomayor, M.A.O.: A Study in Game-theoretic Modeling and Analysis. Cambridge University Press, Cambridge (2006)

Shoham, Y.: Computer science and game theory. Commun. ACM **51**(8), 74–79 (2008)

Strogatz, S.H.: Exploring complex networks. Nature **410**(2), 24–27 (2001)

Wang, X., Jiang, Y.: The influence of the randomness on average path length. Adv. Mat. Res. **87**(19), 198701 (2011)

Author Index

Printed in the United States
By Bookmasters